# LIGHT AND ITS USES

THE AMATEUR SCIENTIST

*Readings from*

**SCIENTIFIC
AMERICAN**

# /LIGHT AND ITS USES/

## MAKING AND USING LASERS, HOLOGRAMS, INTERFEROMETERS, AND INSTRUMENTS OF DISPERSION

With Introductions by
**Jearl Walker**
*Cleveland State University*

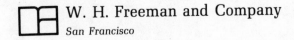 W. H. Freeman and Company
*San Francisco*

**Library of Congress Cataloging in Publication Data**

Main entry under title:

Light and its uses.

    At head of title: The amateur scientist.
    Includes bibliographies and index.
    1.  Lasers—Addresses, essays, lectures.
2.  Holography—Addresses, essays, lectures.
3.  Interferometer—Addresses, essays, lectures.
4.  Spectrum analysis—Instruments—Addresses, essays,
lectures.  I.  Walker, Jearl, 1945-        II.  Scientific
American.
TA1688.L53        621.36             79-27551
ISBN 0-7167-1184-2
ISBN 0-7167-1185-0 pbk.

Printed in the United States of America
9 8 7 6 5 4 3 2

# CONTENTS

# PREFACE

This book is a collection of articles dealing with the optics and construction of lasers, holograms, interferometers, and several types of instruments designed to disperse light into its constituent wavelengths. Most of the articles were written by C. L. "Red" Stong, who conducted "The Amateur Scientist" department of *Scientific American* from 1957 until February 1976. More than a collection, this book is a tribute to Red Stong. He left his imprint on countless thousands of amateur scientists who grew up reading his material and wishing they could perform all of the fascinating experiments he described.

Red Stong was a master of science writing. In his articles, science was intriguing and yet within the grasp of his readers. Stong could reach into the most exotic areas of contemporary science, fashion an instrument, take a few measurements, and then convince us that we could do the same. Even years after his articles were first published, hundreds of his projects are imitated and exhibited in regional and international science fairs. No one has ever had such a strong and lasting influence on students in experimental science as Red Stong.

In the early days of science, scientists were probably very much like Stong. They would tinker with their instruments, peer through lenses, slice white light into its colors, and stay up all night following the stars across the sky. A lot of science has moved on to grander projects, and by doing so it has left much room for the tinkerers. Stong's work is the perfect introduction to science for tinkerers because he was a curious blend of the old and the new in science. On the one hand, he was the experimenter of the classic sort, redoing the important experiments of the late nineteenth and early twentieth centuries. On the other hand, he was at the forefront of the most recent developments in science and boldly led us into some of the important experiments of the mid-twentieth century. When lasers first appeared, they were exotic and almost awesome. When holograms were first invented, they were almost magical. However, within a few years, Stong had cleared out the mystery surrounding both of them and had convinced us that we could build our own lasers and make our own holograms.

The articles collected in this book provide a bridge from the old to the new in the study of light and its uses. In these pages are descriptions of the spectroscopes that were built during the last century to investigate the nature of light and light-emitting bodies. With the spectroscope came the observation that the light emitted by atoms did not fall uniformly across the visible range, but instead displayed mysterious bright lines corresponding to certain discrete wavelengths that atoms emit. This observation foreshadowed the development of quantum mechanics just after the turn of the century. Within the framework of that new physics of atoms, electrons, and photons, the process of stimulated emission was soon discovered. By midcentury the first successful harnessing of the stimulated emission of light was achieved with the laser, and a new experimental and technical era began. The optical instruments whose constructions are described in this book transformed the classical concepts of light and matter into our present concepts and shaped our modern understanding of the atomic nature of matter.

*February 1980*                                                            Jearl Walker

## Note on Supplies

*February 1980*

Please realize that the costs and supply sources given in these articles, especially those dealing with laser components, are now out of date. Some of the supplies can still be purchased from the better-known scientific supply houses, such as Edmund Scientific Company (6982 Edscorp Building, Barrington, NJ 08007). For the more specialized components, you will have to contact companies dealing with specialized products for optics work. Such companies advertise monthly in two magazines, *Laser Focus* and *EOSD* [*Electro-Optical Systems Design*], which are devoted to laser technology. Both magazines publish a yearly listing of all the major suppliers of lasers and laser-related equipment. However, these magazines are normally sent only to professionals working with lasers. To find copies of these magazines (and to ask for advice in general), you might inquire at the physics or laser-engineering department in a local university or at an engineering firm dealing in optics.

# LASERS

I

# LASERS

## INTRODUCTION

Laser technology is rapidly developing, even to the point where the ray guns of science fiction stories no longer appear so fantastic. In the following articles Red Stong describes the plans to make virtually all of the different types of lasers known to operate in or near the visible range. In his basement laboratory Stong built three gas lasers, one liquid laser, and one solid-state laser. These lasers included both pulsed and continuous wave (CW) varieties. Most of Stong's lasers emitted only at a single frequency, but one type emitted at several frequencies simultaneously, and the liquid laser was tunable across a band of frequences.

The basic conditions required for laser action are the same for all types of lasers. Atoms (or, in some cases, molecules) must first be raised to excited energy states from which they subsequently fall to lower energy states, emitting light in the process. It is the special way in which the atoms are stimulated to drop to lower states that characterizes lasers.

When an atom is excited, it temporarily occupies a higher energy level. The atom then deexcites either to the lowest energy level available ("ground state") or to some other state below its present level. Although when the deexcitation occurs is unpredictable, the average time taken by a collection of atoms can be predicted. A laser employs a substance that has an excited state at which a collection of atoms remains for a relatively long time.

When an atom deexcites, it emits light having a certain energy, wavelength, and frequency. The light can be modeled as being a particle (in which case it is called a photon) or as being a wave. If the light emitted by one atom happens to encounter another atom still in the excited state formerly occupied by the first atom, then the passage of the light can force the second atom to deexcite immediately rather than spontaneously at some later time. The light emitted by the second atom then contributes additional light identical to the light forcing the deexcitation. Because the electric field of the passing light stimulates the emission from the second atom, the deexcitation is called stimulated emission. A light bulb emits light because of spontaneous emission from atoms in its heated filament, whereas a laser emits light because of stimulated emission from its atoms.

After an atom undergoes stimulated emission, the two light waves (the one forcing the emission and the one resulting from it) are said to be coherent for three reasons. They have (in the terminology of the wave model) the same wavelength, frequency, and energy. They are in step with each other; that is, the peaks of one wave are aligned with the peaks of the other wave. They are traveling in the same direction. Because of this coherence, the two light waves add to give a net wave whose intensity is larger than the intensity of the wave whose passage caused the stimulated emission. The goal of a laser is to have such stimulated emission repeated a large number of times. Since all the light waves that are produced are coherent with one another, they add to give a huge wave with a hugh intensity. Lasers are very bright because the deexcitation process results in such an addition of light waves. A light bulb with the same number of emitting atoms is considerably less bright because the waves lack coherence and thus do not add to give a huge wave.

Atoms in a laser can be excited either by absorption of light or by collision. Absorption of light can be accomplished by flooding the laser tube with white light so that the atoms absorb their characteristic wavelengths and are excited to upper energy levels. Eventually the atoms will deexcite by spontaneous emission, and some of the atoms will reach the energy level involved in the laser emission. Pausing there for a relatively long time, they can be stimulated to emit photons. An avalanche of photons can be triggered by a single photon emitted spontaneously by one of these atoms because a photon traveling down the length of the tube will provoke stimulated emission of many other photons that in turn also travel down the tube, stimulating emission of still more photons.

Atoms can also become excited through collision. For example, an electrical discharge can send electrons streaming through a laser tube and crashing into the atoms, thereby exciting them. In some lasers these atoms participate in the stimulated emission; in others they merely transfer their energy to another species of atom that then participates in the stimulated emission. The helium-neon laser is an important example of the latter type. The electrical discharge excites the less massive helium atoms that then excite the more massive neon atoms that, because of their relatively large mass, could not have been efficiently excited directly by the discharge.

Part of the stream of stimulated photons is reflected by mirrors or other reflecting surfaces in the laser. The reflected stream stimulates more photons. The rest of the light, which may be just a tiny fraction of the total light, is allowed to leave, forming the laser beam. The optical arrangement of the laser tube with an external mirror at each end is called a resonant cavity because the light is built up inside the cavity in much the same way that sound waves can be built up on a guitar string to produce a wave of large amplitude.

The dye laser described by Stong in one of the following articles is unusual in that it can operate without one of its mirrors. If enough energy is sent from a flashlamp through the tube of dye, the dye may undergo stimulated emission without either mirror in place. This phenomenon is called superradiance. Such an emission can occur when the increase in stimulated photons is so large along the tube that no feedback of the beam is necessary.

The dye laser is excited with a flashlamp whose light is focused into the dye cell by a reflecting elliptical cavity. You might be able to substitute a simple piece of aluminum foil wrapped around both the flashlamp and the dye cell. (Do not allow the two to touch, however.) The dye could also be excited with an ultraviolet laser such as the nitrogen laser described in one of Stong's articles below. The ultraviolet light must be focused down into the dye cell to optically excite the dye molecules. The lens and the dye cell should be made of quartz in order to avoid premature absorption of the ultraviolet light.

Another interesting feature of the dye laser is that its output is tunable. For a given dye the laser will emit simultaneously over a relatively broad band of wavelengths instead of at just a single wavelength as is the case for the helium-neon laser. If a dispersing element is placed in the laser cavity, between the dye tube and one of the mirrors, the range of wavelengths over which the laser operates can be varied within a large band of wavelengths. Stong suggests that a diffraction grating be used for the dispersion. One or more prisms will also serve, although not as well. By rotating either a grating or a prism to send light back through the tube, you can change the wavelength and color of the light emitted by the laser.

Why does the dye emit stimulated radiation over a range of wavelengths and the helium-neon at a single wavelength? The energy levels of the dye molecules come in bands that are relatively wide compared to the energy levels of single atoms such as neon. As the dye molecules are excited to an upper band, they populate it throughout the range of energies within that band. When the molecules engage in stimulated emission, they will emit a range of wavelengths because of the variety of energies they initially have. (However, if a molecule is to be coaxed into stimulated emission, it still must be passed by a photon having the same energy and wavelength as the stimulated photon will have. Thus the coaxing process results in a single wavelength emission for both types of lasers.)

Another laser described in one of the following articles is the infrared diode laser. This type of solid-state device will soon be seen nationwide as part of an alternative to the conventional telephone system. Telephone conversations will be converted to pulsed light signals that are transmitted through optical fibers and then uncoded at the other end of the line. Solid-state lasers will convert the original signal into light pulses and will also boost the signal strength every 14 kilometers or so. The new system will not only eliminate the need for costly copper, but will provide more transmission capabilities than the present system of cables.

# Helium-Neon Laser

## A helium-neon laser built in the home by an amateur

*September 1964*

Few devices in the modern physics laboratory present a more deceptive appearance of simplicity than the helium-neon laser, a device with many exciting prospects [see "Advances in Optical Masers," by Arthur L. Schawlow; SCIENTIFIC AMERICAN, July, 1963]. The apparatus seems to consist merely of a gas-discharge tube that looks much like the letter "I" in a neon sign; at the ends of the tube are flat windows that face a pair of small mirrors. Yet when power is applied, the device emits as many as six separate beams of intense coherent light.

What accounts for this remarkable performance? You can discover the answer by the pleasant if somewhat slow method of undertaking to build a laser at home, as I did earlier this year. Not only will you learn what makes the laser tick but also, as a bonus, you may encounter some fascinating properties of light that you have previously overlooked. If you are as inexperienced as I was, however, you may not find the project as easy or as inexpensive as some that have been discussed in these columns. I can promise that it will exercise your talents for such diverse arts as blowing glass, fabricating small parts, maintaining scrupulous cleanliness in the workshop and operating a high-vacuum system. The cost will vary inversely in proportion to your capacity for improvisation, but you can expect it to exceed $100.

The gas laser requires special structures, the need for which arises because the device is an extremely poor amplifier, at least by electronic standards. In the visible region of the spectrum it usually has a maximum gain of less than 10 percent. Much of the input power is wasted by excited atoms of ionized gas that emit light in random directions. Some emission, however, travels along the axis of the gas-discharge tube and is reflected back and forth between the mirrors. During each transit this oscillating light stimulates still other excited atoms to emit energy that falls into lockstep with the same waves that triggered the emission. The stimulated emission thus increases the intensity of the light,

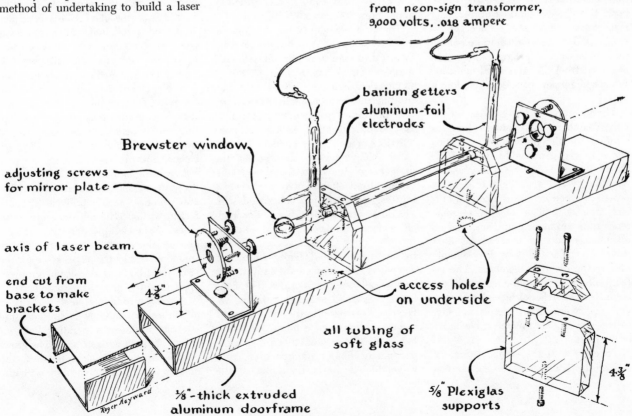

*A helium-neon laser designed for amateur construction*

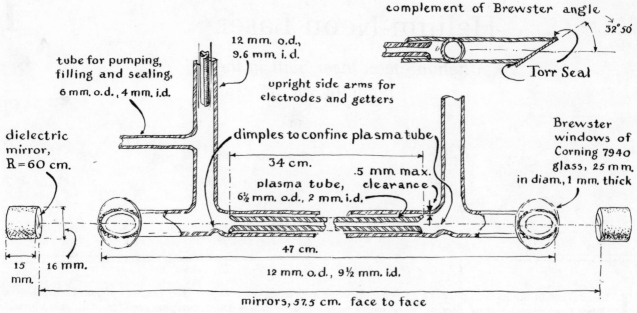

tube for pumping,
filling and sealing,
6 mm. o.d., 4 mm. i.d.

12 mm. o.d.,
9.6 mm. i.d.

upright side arms for
electrodes and getters

complement of Brewster angle

32°50'

Torr Seal

dielectric
mirror,
R = 60 cm.

dimples to confine plasma tube

34 cm.

plasma tube,
6½ mm. o.d., 2 mm. i.d.

.5 mm. max.
clearance

Brewster
windows of
Corning 7940
glass, 25 mm.
in diam., 1 mm. thick

15
mm.

16 mm.

47 cm.

12 mm. o.d., 9½ mm. i.d.

mirrors, 57.5 cm. face to face

*Details of the laser tube assembly*

but only by a few percent on each pass.

During each transit the light must make its way through the end windows to the mirrors and back again. These surfaces are obstacles that can introduce major losses both by absorption and by reflection, as is evident from what happens with an ordinary glass window. When light falls at right angles on such a window, about 4 percent of the energy is reflected back to the source by each of the window's two surfaces. A lesser amount is also absorbed, being transformed into heat by the glass. In an imperfectly aligned window of the laser these losses combine to reduce the intensity of a ray more than 16 percent in the course of a single pass, more than is gained from the stimulated emission. Perfect alignment of a laser window is impossible.

The inventors of the gas laser at the Bell Telephone Laboratories found a solution for the problem of reflection losses in the century-old work of Sir David Brewster, the Scottish physicist who discovered that light is strongly polarized when it falls at a certain critical angle on a sheet of glass or some other transparent medium, and that no reflection occurs in the case of light waves so polarized that they vibrate only in the plane of incidence. The effect is observed when the tangent of the angle between the surface of the window and a line drawn perpendicularly to the rays equals the refractive index of the glass.

To cope with the second source of loss at the laser windows—the conver-

sion of light into heat—the designers simply substituted quartz for glass. At a wavelength of 6,328 angstrom units, which is the wavelength of the light in a helium-neon laser, the heat loss in quartz is slight compared with that in glass. The refractive index of quartz is 1.54, equivalent to an angle of 57 degrees. Quartz plates installed in the laser at this angle are called Brewster windows. If the Brewster windows at each end of the laser envelope are in the same plane of polarization, the transmission of light through the assembly approaches 100 percent.

The designers found an equally ingenious solution for the problem of losses at the mirrors. Freshly silvered glass reflects as much as 96 percent of the incident light, but as the metal film tarnishes in the atmosphere its performance falls sharply. Aluminum, the next best metallic coating, reflects at most only 92 percent of the incident light.

To achieve higher performance the designers of the laser abandoned conventional reflectors in favor of dielectric mirrors, which are mirrors coated with several nonconducting films. For high efficiency such mirrors depend on interference among the light waves reflected by the multilayered films, which are composed of transparent substances such as sulfides and fluorides. The films can be designed either to suppress reflection, as they do in the familiar coating on the lenses of cameras, or to enhance it. If the refractive index of a transparent film a quarter of a light wave thick is lower than that of the

glass on which it is deposited, a wave of light reflected by the glass arrives at the surface of the film 180 degrees out of step with a wave reflected by the film. The crest of one wave falls in step with the trough of the other and the two cancel. If the refractive index of the film is higher than that of the glass, the crests and troughs of the two combine to increase the reflectivity.

A second film of lower refractive index than the glass, when applied over the first film, reflects waves of opposite phase with respect to those reflected from the glass. The second film is located a full half-wavelength away from the glass, however, which precisely compensates for the difference in phase. The waves therefore again interfere constructively to increase the reflectivity. The application of a third film reduces the reflectivity somewhat but its effect is more than compensated by the fourth film, and so on. The dielectric mirrors used in lasers employ between 13 and 27 films, and the reflectivity of such a mirror approaches 100 percent. Incidentally, the manufacture of dielectric mirrors requires facilities and techniques that are normally beyond the reach of amateurs. These components, like the gas mixture for the laser, must be bought. A list of suppliers appears at the end of this article.

When the combined losses have been minimized by suitable techniques, the intensity of the reverberating light increases, but not without limit. The growing electromagnetic field between the mirrors interferes increasingly with the number of excited atoms that respond to the influence of the field and so are stimulated to emit energy. A delicate balance is eventually achieved between the losses and the gain. At this point energy gained by stimulated emission precisely equals the combined losses, including the portion that escapes from the apparatus in the output beams.

In addition to these losses, a falling off in gain can occur inside the tube assembly. The assembly consists of a glass envelope that supports the windows, a pair of electrodes and the plasma tube— a slender tube of small bore in which the discharge occurs. Internal losses become serious if the helium-neon mixture is contaminated by even a trace of foreign gases such as oxygen, nitrogen and carbon dioxide, or if the pressure of the gas is not maintained within certain limits. The problem of contamination is met largely by removing from the tube all unwanted gases and all substances that can release vapors. The inner parts of the device must be immaculate. Even the faintest smudge from a fingertip can release an astonishing amount of vapor. The clean tube is partly cleared of unwanted gases by the vacuum pumps. It is evacuated to a pressure of at least $10^{-5}$ torr. (A torr is the pressure that will support a column of mercury one millimeter in height.) The remaining contamination is then immobilized by firing a "getter," an electrically heated crucible inside the tube from which vaporized barium condenses on the glass walls. The barium unites chemically with most elements other than the inert gases.

The optimum pressure at which stimulated emission occurs varies inversely with the diameter of the gas-discharge tube. The pressure in torrs is equal to 3.6 divided by the inside diameter of the tube in millimeters. The laser will continue to operate at diminishing intensity up to about twice the optimum pressure and down to about half of it. Advantage is taken of this fact to extend the service life of laser tubes by overfilling them by 50 percent, because for reasons not fully understood the gas pressure drops slowly as the laser operates. Lasers will operate best on gas ratios of seven parts of helium to one of neon, but the tubes can be filled with a nine-to-one mixture.

The range of tube diameters that can be used in the laser is restricted by the nature of ionized gas and by mechanical considerations. At pressures much lower than .5 torr the electrons acquire enough energy to damage the glass envelope by impact and to erode the electrodes. The metal vaporized in this way condenses on the envelope and in the process buries gas atoms, lowering the pressure still more. Thus a runaway effect develops that causes the tube to fail. An envelope that contains a plasma tube six millimeters in diameter ought to be filled, according to the formula given above, to a pressure of 3.6/6, or .6 torr. With a laser of the type I built the life of a tube filled to this marginal pressure would be impractically short. I use a plasma tube with a diameter of two millimeters, and I overfill it to a pressure of 2.7 torrs. Tube diameters of less than about one millimeter become awkward to align and have other drawbacks.

In general the output of a laser increases in proportion to the product of the length times the diameter of the plasma tube. This suggests that long tubes are more powerful than short ones. Again there is a catch. Short tubes operate readily at 6,328 angstroms, whereas those a meter or more in length tend to function in the infrared region instead of in the visible part of the spectrum. I am told that long tubes can be forced to work in the visible range by the strategic placement of magnets along the tube, but I have not tried the experiment. Tubes ranging from 15 to 40 centimeters in length appear to work quite adequately in the visible region. Mine measures 34 centimeters.

The mirror system functions somewhat like a resonator. It can consist of various combinations of spherical and flat mirrors. A system that is easy to adjust and to maintain in adjustment employs a pair of facing spherical mirrors separated by slightly less than one

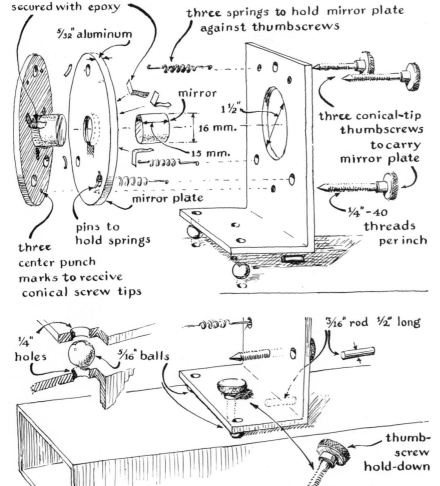

three watch-spring clips secured with epoxy

$5/32$" aluminum

three springs to hold mirror plate against thumbscrews

mirror

$1\frac{1}{2}$"

16 mm.

15 mm.

three conical-tip thumbscrews to carry mirror plate

mirror plate

pins to hold springs

three center punch marks to receive conical screw tips

$\frac{1}{4}$"-40 threads per inch

$\frac{1}{4}$" holes

$5/16$" balls

$3/16$" rod $\frac{1}{2}$" long

thumb-screw hold-down

$\frac{1}{8}$" wall

aluminum doorframe

*Parts and arrangement of fixture to support mirror cell*

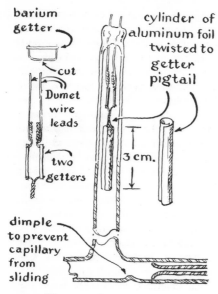

barium
getter

cut

Dumet
wire
leads

two
getters

cylinder of
aluminum foil
twisted to
getter
pigtail

3 cm.

dimple
to prevent
capillary
from
sliding

*The getter-electrode assembly*

radius. My mirrors were figured to a radius of 60 centimeters. The spherical surfaces have a separation of 57.5 centimeters. The mirrors are mounted in easily adjustable cells supported by fixtures that can be removed and returned to the base without disturbing the alignment of the mirrors with respect to the axis of the tube. All essential dimensions of the laser's hardware, of the light projector used for aligning the mirrors and of the miter box for sawing the envelope at the Brewster angle are specified in the accompanying illustrations. Alternate design schemes are possible and perhaps desirable. The reader is encouraged to improvise.

Gas lasers can be energized by alternating current of either high or low frequency and by direct current. The direct-current types that employ heated electrodes have a long service life. Making the heated electrodes, however, is an intricate job that I am reluctant to undertake now. My laser is equipped with cold electrodes made of aluminum in the form of small cylinders and is energized by a conventional neon-sign transformer of the constant-current type. The primary voltage is controlled by a Variac, a variable-voltage transformer. When power is applied to the primary of the neon-sign transformer, the secondary winding maintains 18 milliamperes through the load at a maximum potential of 9,000 volts. Experiment demonstrated that the laser beam reaches maximum intensity when 85 volts is applied to the primary of the neon-sign transformer.

I began building the laser by making the getter-electrode assemblies. The getter is a small metallic trough, filled with barium, that is connected at the ends to a loop of wire. When it is installed inside an evacuated glass envelope, the loop can be coupled to a high-frequency electromagnetic field and heated by induced current to vaporize the barium. That is the conventional procedure. I do not own an induction-heating apparatus of this type, so I cut the loops of two getters, spliced the pair in series by means of a pigtail joint and hooked a pair of Dumet leads to the free ends as shown in the accompanying illustration [*top of this page*]. Dumet is a spe-

cial alloy wire that seals readily to soft glass. I fire the getters by hooking the Dumet leads to a six-volt transformer that is energized through the Variac. The assembly becomes red-hot when a potential of about two volts is applied to the leads and yellow-hot at three volts. At that temperature the barium vaporizes and condenses in the form of a dime-sized black film on the glass surface of the evacuated tube. The pair of getters draws seven amperes at three volts. I apply the heating current slowly, allow about a minute for the temperature to rise to yellow heat and then switch off the power promptly when the film of condensed barium becomes almost opaque. The units contain enough barium for about five such firings. It is easy to make the mistake of increasing the power too quickly and exploding the wire.

The pigtail splice between the two getters serves as the support for the cylindrical electrode: a ribbon of clean aluminum foil in the form of a single, slightly overlapped turn. I make it by winding the foil around the end of a six-millimeter glass rod. One end of the resulting cylinder is then twisted around the getter pigtail. Before touching the getters and foil thoroughly clean your hands, as well as any tools, with carbon tetrachloride. Take the foil from the inside of a new roll.

The getter-electrode assemblies are installed in two short lengths of glass tubing that become side arms of the glass envelope. One side arm is equipped with a smaller tube for exhausting and backfilling the envelope. I then flare the ends of the plasma tube alternately by blowing a small bulb on one end, exploding the softened bulb and shrinking its circular edge in the fire until, by trial and error, the flared end fits the inside of the envelope to within a half-millimeter, or closer if possible. A dimple is next sucked in the envelope as a stop for the plasma tube. The outside of the plasma tube is cleaned with fuming nitric acid, rinsed with distilled water, dried in the flame and slipped into the envelope against the dimple stop. The second dimple is then made in the envelope to secure the tube loosely. Next the side arms containing the getter-electrode assemblies are joined to the envelope. Incidentally, if you have the glasswork made commercially, you will miss a lot of fun. Working with hot glass, particularly in the case of simple apparatus such as this, is not nearly so difficult as is commonly supposed. I discussed glassblowing in this department in May.

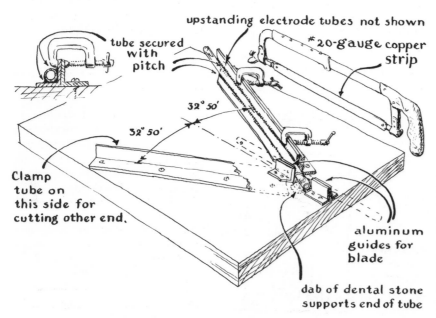

upstanding electrode tubes not shown

tube secured
with
pitch

#20-gauge copper
strip

32° 50'

32° 50'

Clamp
tube on
this side for
cutting other end.

aluminum
guides for
blade

dab of dental stone
supports end of tube

*Miter box for sawing glass at the Brewster angle*

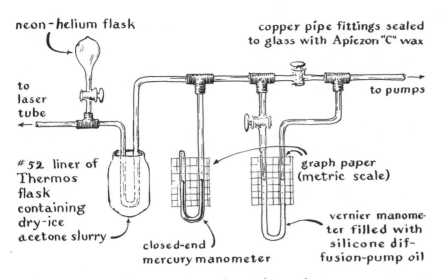

neon-helium flask

to laser tube

#52 liner of Thermos flask containing dry-ice acetone slurry

copper pipe fittings sealed to glass with Apiezon "C" wax

to pumps

graph paper (metric scale)

closed-end mercury manometer

vernier manometer filled with silicone diffusion-pump oil

*Schematic arrangement of system for metering gas*

the pitch that serves as cement. Take out the cotton wads. Clean the soiled inner ends of the envelope with a swab of cotton moistened first with acetone and then with distilled water. Dry the cleaned ends in a gas flame. Remove the pitch with turpentine.

To install the Brewster windows clamp the tube in an apparatus stand or a comparable fixture, connect the side-arm tubing to the mechanical pump of the vacuum system and start the pump. Simultaneously place the cleaned Brewster windows flat against the cut ends. Suction will hold them in place. Using a toothpick as a spatula, apply a thin layer of Torr Seal epoxy cement to the exterior of the joint between the windows and the tube ends. (Torr Seal is manufactured by Varian Associates,

The glass construction is completed by sawing the ends of the tube at the Brewster angle. If you have access to a diamond saw equipped with an accurate fence, the job will take about three minutes. If not, build the miter box shown in the bottom illustration on page 10. With this device the cuts will require about 10 minutes each. Keep plenty of abrasive slurry on the copper blade, let the weight of the saw do the work, use about 60 strokes a minute and take it easy as the saw cuts through the glass. Before making the cuts, plug the ends of the tube with wads of clean absorbent cotton and cement the glass to the right-angled aluminum holder.

The cut ends must be lapped to make a vacuum-tight fit with the Brewster windows. Begin the lapping operation with approximately 400-mesh grit, either Alundum or Carborundum, using a two-inch square of quarter-inch plate glass as the tool. When the pits that were made in the glass by the saw have *almost* disappeared, shift to 600-mesh grit and continue lapping until pits left by the 400-mesh grit have *almost* disappeared. Examine the work through a magnifying glass. The size and number of the remaining pits indicate where the glass is being removed and by how much. If pits disappear slower at one point than at another, exert more grinding pressure on the plate glass above that region. The idea is to remove glass evenly over the entire area of the cut end, thus preserving the Brewster angle. The job requires about 20 minutes. When the lapping is finished, remove the tube from the fixture by warming

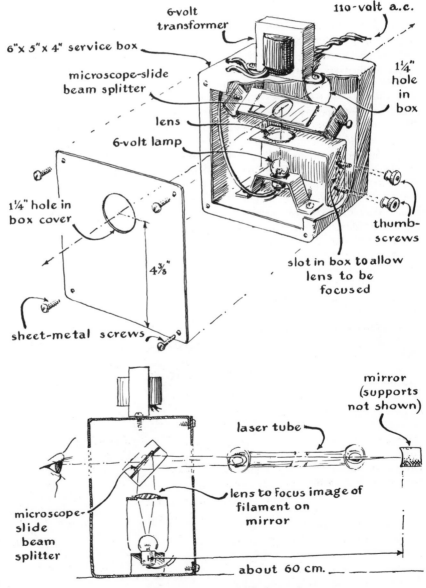

6-volt transformer

110-volt a.c.

6"x 5"x 4" service box

microscope-slide beam splitter

lens

6-volt lamp

1¼" hole in box

1¼" hole in box cover

4⅜"

sheet-metal screws

thumb-screws

slot in box to allow lens to be focused

mirror (supports not shown)

laser tube

microscope-slide beam splitter

lens to focus image of filament on mirror

about 60 cm.

*Light projector for aligning the laser mirrors*

Palo Alto, Calif.) Watch the inner surface of the windows carefully for any trace of vapor from the cement that may be sucked into the tube. If the least trace of vapor appears, the lapping job is defective and must be corrected. No cement should seep through the joint. The curing time of the cement can be shortened by applying heat to the joints by means of a pair of 150-watt incandescent lamps placed about 15 centimeters away. Recommended curing temperatures are specified on the package by the manufacturer. When the joints have cured, apply a coat of Dow Corning 806A silicone resin over the cement, all glass joints and the Dumet seals.

Next fire up your vacuum system and connect the tube. The system should be equipped with at least two manometers: one a closed-end mercury type and the other a conventional manometer half-filled with high-grade oil of the kind used in a diffusion vacuum pump. Scales for the manometers can be made of graph paper calibrated in millimeters [see illustration on page 11]. The oil manometer acts as a vernier gauge, the pressure readings in millimeters being converted to torrs by dividing the specific gravity of mercury (approximately 13.5) by the specific gravity of the oil and then dividing the difference in the height of the oil in the two arms of the manometer, in millimeters, by the quotient. For example, if the ratio of the specific gravity of the mercury to that of the oil is 16, a manometer reading of 32 millimeters—the difference in the height of the oil in the two arms—indicates a pressure of 32/16, or 2 torrs. You can find the specific gravity of the oil with sufficient accuracy by weighing 10 milliliters and dividing the weight (in grams) by 10.

Your vacuum system should also contain a McLeod gauge, primarily for ensuring that the system pumps to the required $10^{-5}$ torr. If you do not own a vacuum system, you may be interested in building the inexpensive one described in the Scientific American Book of Projects for the Amateur Scientist (published by Simon and Schuster in 1960).

My system uses a two-stage mercury-jet pump. For valves the system has inexpensive glass stopcocks. I lapped them with 600-mesh grit to make a vacuum-tight fit and lubricated the stoppers with a thin film of high-vacuum grease. One stopper was modified to function as a leak valve for admitting the required minute volume of helium-neon gas to the laser tube. Deep

scratches about six millimeters long that tapered upward to the surface were cut counterclockwise from each end of the hole in the stopper. The tips of the scratches conduct gas at a convenient rate when they engage the openings of the barrel.

You will receive the helium-neon gas in an all-glass container. The outlet tube contains an easily broken seal. To tap the supply clamp the flask temporarily to a support, with the tube pointing downward. Insert a loosely fitting glass marble or a short length of heavy glass rod in the tube, join the leak valve to the tube with Apiezon "C" wax and connect the valve to the vacuum system by 50 centimeters or so of copper capillary of the type used in the thermostats of gas stoves. Open the leak valve fully and exhaust the tube to $10^{-5}$ torr, or lower if possible. Then close the valve and invert the flask quickly so that the marble falls and breaks the tip of the seal.

The laser tube is next exhausted, cleaned by discharge bombardment, filled to the required pressure and sealed off. For support the tube can be assembled to the base, in which case the laser can be tested before the seal-off.

Pump the tube down to $10^{-5}$ torr or lower, then backfill with helium-neon to a pressure of approximately 5 torrs. Connect and switch on the 9,000-volt transformer. The tube will fill with colored plasma—reds, greens, blues—and may get quite hot, on the order of 100 degrees centigrade or more. After five minutes switch off the power, repump the tube and repeat the cycle. Continue the procedure until the reddish color

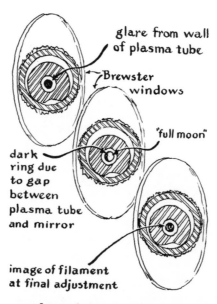

glare from wall
of plasma tube

Brewster
windows

"full moon"

dark
ring due
to gap
between
plasma tube
and mirror

image of filament
at final adjustment

*Images during mirror alignment*

predominates, tinged only slightly by blue. This may require four but probably not more than six cycles. Before pumping down at the end of the last bombardment heat the getters to dull red. The tube should be positioned so that both getter assemblies are suspended vertically by their leads, otherwise they may sag enough to touch the glass when heat softens the metal. After about three minutes advance the Variac gradually to fire the getters. Watch the glass wall adjacent to the getter assembly as the temperature of the metal approaches yellow heat. A sooty film will appear. As the film reaches opaqueness switch off the power. Having pumped down, refill the tube to a pressure of *not more* than 2.7 torrs (assuming a plasma tube of two-millimeter bore). Energize the tube again. The plasma should now appear solid reddish-orange. When placed between a pair of properly adjusted mirrors, the tube will now function.

To adjust the mirrors, first remove either of the cell fixtures from the base and replace it with the light projector. Adjust the lens of the projector until the image of the filament is in focus on the front surface of the distant mirror. Then, using your hand as a screen at the near end of the base, locate the beam reflected by the mirror and adjust the cell to center the beam on the distant end of the plasma tube. Look through the projector's beam splitter (the microscope slide) into the bore of the plasma tube; in the center of the glare that is reflected by the inner wall you may observe a minute disk of light that resembles the full moon surrounded by a thin, dark ring. The "moon" is the reflection of scattered light from the projector. Adjust the cell in any direction that causes the disk to brighten. Ultimately it will become dazzling as the reflected image of the filament comes into view [see illustration at left]. This completes the adjustment of the first mirror. Remove the adjusted cell fixture carefully and similarly adjust the second one. When both adjusted mirrors have been assembled to the base, connect the neon-sign transformer to the tube and switch on the power. Usually it is now necessary to "fiddle with the screws." Just rock the adjustment screws back and forth a degree or so, one after the other. Suddenly the beam will appear—you have a laser! Seal off the tube.

The beam, when it is directed against a screen, will doubtless appear in the form of a symmetrical pattern of dazzling red spots—perhaps only a pair, maybe a rosette of eight spots or some other geometric design. The pattern can be greatly altered and perhaps intensified simply by fiddling with the adjustment screws. Each pattern of spots results from a unique set of paths taken by rays that oscillate between the mirrors. The various patterns of vibration are known as "modes." Observe too the scintillating granularity of the light. According to E. I. Gordon, a physicist at the Bell Telephone Laboratories, this striking effect actually arises in the eye or any other image-forming device, such as a camera. Each dazzling speck, he explains, marks a point of constructive interference between coherent diffraction patterns; the size of the point is determined by the diameter of the pupil. Phenomena such as the granularity suggest many fascinating experiments that can be made with laser light. Some will be discussed from time to time in this department.

Now, a word of warning: The laser is a hazardous apparatus. *Never look directly into the beam.* Coherent light of this intensity can damage the retina permanently and may even cause blindness. When other people are in the room, block off all beams at points close to the apparatus: the two beams from the ends and the remaining four of lesser intensity that come off the Brewster windows. The 9,000-volt output of the neon-sign transformer is lethal. Insulate the full length of the leads with abutting pieces of glass tubing. Never touch the terminals of the laser when the transformer is plugged into the power line even if you are certain that the switch is off.

Parts and materials for constructing the laser can be procured from the following suppliers:

Perkin-Elmer Corporation, Electro-Optical Division, Norwalk, Conn. To encourage student experimentation in optics this firm has developed special dielectric mirrors of adequate quality for apparatus of the type described in this article.

Edmund Scientific Co., Barrington, N.J. This organization stocks lenses and related materials.

Morris and Lee, 1685 Elmwood Avenue, Buffalo, N.Y. 14207. Air pumps, pressure gauges, valves and accessories for vacuum systems are made specially for amateurs by this organization.

Henry Prescott, Main Street, Northfield, Mass. This supplier specializes in all materials required for constructing and experimenting with the laser, including dielectric mirrors, vacuum systems, helium-neon gas, glass components, getters and related essentials.

# 2

# More on the Helium-Neon Laser

*Increasing the life of the amplifier tube at modest cost*

December 1965

More can be said about the helium-neon laser previously described here (September 1964). Lasers of this type are in effect electromagnetic oscillators. Essentially they consist of an amplifier tube and a resonant cavity. They emit continuous beams of intense coherent light at 6,328 angstrom units and open new experimental opportunities for amateurs in several disciplines. My own interest in recent months has centered on improving the performance of the apparatus—particularly by increasing the useful life of the amplifier tube—without substantially increasing the cost over that of the unit I had originally constructed.

The amplifying portion of my apparatus consists of a straight gas-discharge tube, equipped with cold electrodes, that contains a mixture of helium and neon in the ratio of seven to one at a pressure of 1.8 to 2.7 torr, depending on the diameter of the tube. The ends of the tube are closed by plane windows set at an angle equal to the trigonometric cotangent of the index of refraction of the window material, which may consist of any clear pure glass or of fused quartz. The windows are attached to the tube by epoxy cement. The resonant cavity consists of a pair of dielectric mirrors of spherical curvature mounted to face each other at a distance equal to approximately 95 percent of their radius of curvature and

adjusted so that their optical axes coincide. The amplifier tube is positioned coaxially between the mirrors.

When an electric field of sufficient potential to ionize the gas is applied to the electrodes, excited atoms of helium collide with and transfer energy to the neon atoms, raising the neon atoms to one or another of their higher energy levels. Subsequently the neon atoms spontaneously drop to one or another of the lower energy levels they naturally occupy and simultaneously emit light of the wavelength that is characteristic of the energy released. Some transitions occur between the levels that give rise to emission at the wavelength of 6,328 angstroms.

Some photons of this wavelength are emitted along the axis of the tube. The energy then oscillates between the mirrors. During each transit through the tube this oscillating energy stimulates neon atoms that happen to occupy the appropriate energy level to drop to the appropriate lower level and thus contribute their energy to the resonator.

In lasers of this type the intensity of the oscillating light can increase as much as 5 percent during each transit through the amplifying tube. Although more than 500 million transits are made each second, energy stored in the resonator does not increase without limit. The efficiency of the amplifying action decreases as the stored energy in-

creases. Efficiency is also impaired by the presence of impurities in the gases, departures from optimum gas pressure and changes in the ratio of helium to neon. In addition the stored energy is dissipated in various ways. Some is scattered by dust and imperfections on the surfaces of the mirrors and windows. Another portion is diverted by reflection from the surfaces of the windows. Ultimately less than two-tenths of 1 percent seeps through the mirrors. This small fraction constitutes the useful output of the laser.

The amplifying tube of the first apparatus I constructed operated only 15 hours before failing. That the laser worked at all was gratifying, of course, but I felt that I was entitled to a longer run for my pains. The immediate cause of failure was "sputtering," which is the erosion of the metal electrodes by the electrified gases. Metal thus eroded collects in part as a film on the glass walls and lowers the pressure of the gas by burying atoms of helium and neon in the debris. The lowered pressure accelerates the phenomenon, which soon leads to the destruction of the tube.

Some metals tend to sputter more readily than others. The destructive action can be minimized by coating electrodes with a metal such as barium that increases conductivity in the vicinity of the electrode, thereby lowering the velocity of the ions in the plasma and the consequences of their impact on the metal. Coated electrodes must be purged of the occluded gases when the tube is constructed. This is usually done by operating the tube for a time on current intense enough to heat the metal to redness. I omitted the step during my initial construction in an attempt to minimize the cost of the project. By trial and error I found that electrodes of aluminum foil resist erosion without having to be heated. The oxide coating naturally present on aluminum appears to retard the sputtering action for as long as 50 hours of operation if all other conditions are favorable.

Sputtering appears to be accelerated by a trace of almost any organic vapor. This became evident from the operation of the manometer I used originally for measuring gas pressure. The instrument employed phthalate as the indicating fluid. The vapor passed readily through a trap refrigerated by a slurry of dry ice and acetone and eventually contaminated the entire vacuum system. From then on even aluminum electrodes sputtered severely after an hour of operation, and no tube gave as many as

five hours of service.

The difficulty was overcome by substituting for the manometer a gauge of the McLeod type, which uses mercury as the indicating fluid. The version of the gauge that I made consists of three joined capillary tubes, each of which contains a bulb [*see illustration on this page*]. One bulb terminates in a closed capillary about 7.5 centimeters long. The middle leg contains a small spherical bulb that is connected to the vacuum system by means of a coil of copper tubing. The third leg terminates in a bulb substantially larger than the other two.

The glass structure is supported on a fixture improvised of plywood and pipe fittings that allows the assembly to be rotated through an angle of about 100 degrees for transferring the mercury by gravity from the largest bulb, in which it is stored, to the other two legs of the gauge. Normally the gauge is kept in the standby position with the mercury in the large bulb. To make a reading the assembly is rotated to the upright position. This causes the mercury to run into the other legs. A specimen of the gas under measurement is trapped and compressed in the closed capillary, limiting the height to which the mercury can rise in the tube.

When the system is fully exhausted, mercury rises to the top of the closed capillary, indicating zero pressure. Simultaneously the metal rises to the same level in the bulb of the middle leg as well as in the reservoir. When the system is not fully exhausted, compressed gas prevents the mercury from rising to the end of the closed capillary. The distance between the closed end and the meniscus, or curved top, of the mercury is a measure of the pressure. In effect the gauge acts as a closed-end manometer.

A vertical scale, plotted in torr, is adjacent to the closed capillary, the zero graduation coinciding with the closed end. Graduations representing higher pressures are plotted at appropriate distances below the zero graduation. Only two quantities must be determined to compute the locations of the graduations: (1) the volume of the closed capillary plus the volume of the bulb to which it is attached and the volume of the capillary that connects this bulb to the middle leg of the gauge, (2) the cross-sectional area of the closed capillary.

To measure the volume, first weigh the glassware (to a tenth of a gram), then fill the volume to be determined

with mercury and weigh it again. Subtract the weight of the glass from that of the glass and the mercury combined to determine the net weight of the mercury. Divide the net weight of the mercury by .0135 to determine the volume of the glassware in cubic millimeters. The cross-sectional area of a two-millimeter capillary is approximately three square millimeters.

To determine the distance in millimeters at which mercury will stand below the zero graduation for any pressure first multiply the volume just measured by the given pressure. Then divide this product by the cross-sectional area of the closed capillary. The square root of this quotient is equal to the distance in millimeters. For example, assume that the position of the graduation is desired for a pressure of 1.5 torr, that the volume of the closed capillary and its associated bulb and connecting tubing is 4,500 cubic millimeters and that the cross-sectional area of the closed capillary is three square millimeters. The distance between the zero graduation and the desired graduation is then equal to the square root of $1.5 \times 4,500/3$, or 47.4 millimeters. To compute the entire scale, make a list of selected pressure intervals, such as .1, .5, 1, 1.5 and so on, and do the arithmetic. Plot the resulting distances in millimeters as graduations on a cardboard scale and cement it to the closed capillary with the zero indication adjacent to the closed end. If the volume of the closed capillary, its associated bulb and connecting tubing is approximately 4,500 cubic millimeters, a 75-millimeter scale will span the range of pressure from zero to three torr. Helium-neon lasers operate within this range.

The useful life of the amplifier tube is also reduced by the release of gases naturally present in the metal of the electrodes. Such gases can be dislodged by repeatedly filling the tube with the helium-neon mixture to a pressure of a torr or so and energizing it with an alternating current of 18 milliamperes, the rated output of the neon-sign transformer from which the tube operates. This is a time-consuming business, however. I now use direct current at 120 milliamperes for heating, and thus driving the gas out of, electrodes of metals other than aluminum. The current is supplied by a General Electric constant-current transformer (Model 916Y11) that develops 15,000 volts at 60 milliamperes. The unit is energized from the 120-volt power line through a variable-voltage transformer. The secondary of

3/8" copper capillary connecting to vacuum system

solder

1/4" copper sleeve

Apiezon "C" wax

volume 7 cc.

8.1 cm.

3.75 cm.

O
.01
.1

volume 1 cc.

7.5 cm.

1
1.8 cm.

4.1 cm.

2
2-mm. bore

3

20.6 cm.

5.6 cm.

volume 4.5 cc.

*Arrangement of the McLeod gauge*

the constant-current transformer is tapped at the center of the secondary winding and grounded to the case by the manufacturer. The output of the secondary is converted to direct current by a conventional full-wave rectifier circuit that employs silicon diodes.

To outgas the electrodes the amplifier is first filled with about three torr of the helium-neon mixture. Power is then applied. The discharge current is gradually increased from minimum to 120 milliamperes during an interval of about 30 seconds. Ion bombardment heats the cathode to redness and the released gases change the color of the plasma from red to blue. Power is then shut off and the tube is pumped down. The procedure is repeated until the electrode is fully outgassed, that is, until the plasma retains its red color. The polarity is reversed for similarly processing the second electrode. Getters are assembled in a sidearm that serves as a reservoir for stabilizing the pressure of the gas. The glass of the reservoir and inner wall of the envelope of the amplifier tube is outgassed by initiating a discharge on alternating current at 18 milliamperes between the getters and the electrodes.

I am now experimenting with iron electrodes in the form of a cylinder, the inner wall of which is coated with barium strontium carbonate, and also with similar electrodes of titanium foil. The titanium is coated by evaporating barium inside the cylinder from a conventional getter of the KIC type. Tubes with coated electrodes of iron have operated from 50 hours to more than 100. Thereafter they can be reconditioned. (Coated-iron electrodes, barium getters, McLeod gauges and other supplies for experimenting with helium-neon lasers can be procured from Henry Prescott, 150 Main Street, Northfield, Mass. 02118.)

To open a spent amplifying tube for replacing the gas or for some other repair, crack the filling tube by heating the tip of the glass in a sharp flame. After an hour or two, when the amplifier has reached atmospheric pressure, the cracked tip can be removed. The slow leak prevents an inrush of air from depositing dust on the windows. If electrodes are also replaced, the glass-blower must insert a desiccator in the blow hose to prevent moisture from condensing on the windows. I use an air filter charged with anhydrous calcium sulfate. Smudged windows seriously reduce the output.

A simple procedure for aligning the mirrors of the resonator has been described in the *American Journal of Physics* for March by K. L. Vander Sluis, G. K. Werner, P. M. Griffin, H. W. Morgan, O. B. Rudolph and P. A. Staats. The alignment tool consists of a square of white cardboard of any convenient size that is blackened on one side. A pair of fine diagonal lines are ruled on the white side of the card, which is pierced with a half-millimeter pinhole at the point where the lines intersect. A quarter-inch square of red gelatin (roughly the color of an Eastman Wratten No. 29 filter) is placed over the pinhole on the black side and cemented in place at the edges.

To align the resonator a small sheet of glass, such as a microscope slide, is placed at approximately right angles to the axis of the tube between one mirror and its facing window. This "spoiler" glass prevents the laser from generating a beam. The experimenter now energizes the amplifier tube and, from the blackened side of the card, looks through the pinhole and either of the mirrors into the capillary of the amplifier tube. The position of the eye is changed until the distant end of the capillary appears as a concentric circle inside the larger circle, which is the near end of the capillary. While the eye is in this position the adjustment screws of the mirror cell are manipulated until the image of the crossed lines, which is reflected by the near side of the mirror, is brought pre-

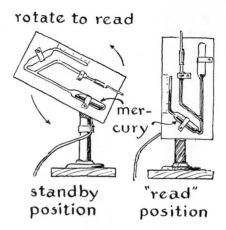

rotate to read

mer-
cury

standby
position

"read"
position

*Reading the gauge*

cisely to the center of the inner circle. The second mirror is then similarly adjusted.

The screws are now carefully rocked a degree or so in each direction. At some point a crescent of bright light will appear at the edge of the inner circle, resembling the rising moon. The screws are then gradually manipulated to produce a full moon. The second mirror is similarly adjusted for the full-moon condition. When the spoiler glass is removed, the apparatus will "lase." The adjustment is now trimmed for maximum output. Caution: *Never adjust the mirrors by this method unless the spoiler glass is in position. The beam may form and permanently damage the experimenter's eye.*

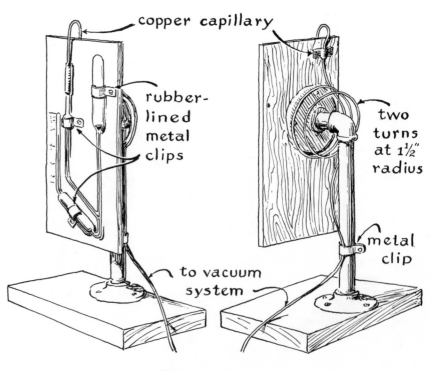

copper capillary

rubber-
lined
metal
clips

two
turns
at 1½"
radius

metal
clip

to vacuum
system

*Mounting of the gauge*

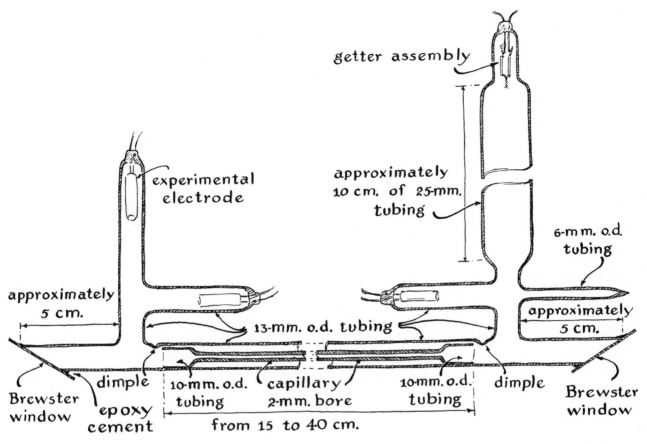

*Arrangement of an experimental helium-neon laser tube*

---

### Note on Cleaning the Mirrors

*November 1966*

Carl Fromer, a high school student in Staten Island, N.Y., has hit on an effective method of cleaning the soft-coated dielectric mirrors used in continuous lasers of the helium-neon type. The coatings accumulate a film of grime after some weeks of exposure to the air. The film seriously reduces their reflectivity. Any chemical treatment that dissolves the grime destroys the coatings.

Fromer removes the grime by electronic bombardment. He inserts the soiled mirrors face up in a simple gas-discharge tube, exhausts the tube to a pressure of about one torr, backfills with argon or with a 7 : 1 helium-neon mixture to a pressure of 10 torr and operates the tube for one minute at a current of 60 milliamperes [*see bottom illustration on this page*]. The mirrors emerge looking like new.

In addition to grime, the coatings are vulnerable to moisture. Hence the mirrors should be stored in a dry place. A good place is an airtight four-ounce jar. Put an ounce of anhydrous calcium sulfate on the bottom and then add a layer of absorbent cotton.

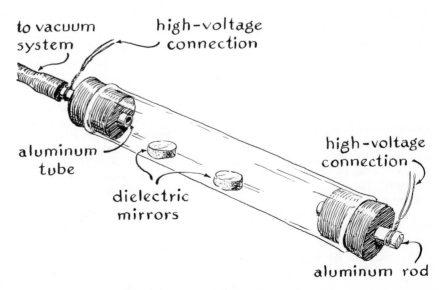

*Carl Fromer's device for cleaning laser mirrors*

# 3

# Argon Ion Laser

*An argon gas laser with outputs*
*at several wavelengths*

February 1969

During the past four years several thousand amateurs have built helium-neon gas lasers of the kind described in this department for September, 1964, December, 1965, and February, 1967. Amateurs have used the lasers, which emit a reddish-orange beam of coherent light at a wavelength of 6,328 angstroms, for such diverse purposes as demonstrating the physical properties of light, testing the optical quality of lenses and mirrors, precisely measuring length and velocity and making holograms. Recently other kinds of lasers that can be built at home have been developed. One of them is the argon gas laser, which emits coherent light of several colors in the green, blue and violet regions of the spectrum and thus greatly enlarges the scope for experimentation.

The argon laser is no more difficult to build than the helium-neon type, although in general the construction of lasers makes a considerably more severe demand on the craftsmanship of the experimenter than most of the projects that have been described in these columns do. Such demands are minimized, however, by an argon laser that has been designed recently for amateur construction by Sylvan Heumann of 410 Eucalyptus Avenue, Hillsborough, Calif. 94010. Heumann writes:

"The argon laser resembles the helium-neon laser in many ways. It consists essentially of a gas-discharge tube about two feet long, the ends of which are closed by a pair of flat windows of fused quartz that face a pair of small dielectric mirrors [*see top illustration on opposite page*]. The tube glows dark blue when the gas is energized by a pulsed electric current of 15 to 20 amperes. Depending on the amount of the current, the energy of the ionized atoms is increased one or more levels above that of the ground state, which is the level of energy that characterizes electrically neutral atoms of argon gas. The electrical discharge is said to pump the atoms to an excited state. After a short time the atoms spontaneously fall to a lower energy state, one at a time, simultaneously emitting a quantum, or pulse, of light. The color of the emitted light varies according to the amount of energy that is liberated during the fall. Quanta of relatively low energy appear red and those of increasingly higher energy appear yellow, green, blue and so on.

"Occasionally a quantum of light that has been liberated spontaneously from an excited atom encounters another energized atom. The resulting interaction may cause the energized atom to fall to a lower energy level and simultaneously liberate a quantum of precisely the same color as the stimulating quantum. This is the phenomenon called stimulated emission. The initial photon causes the excited atom to fall somewhat earlier than it would if it were not disturbed. The two photons merge and proceed through space as a train of coherent, monochromatic light waves.

"The train may encounter a third excited atom and similarly cause it to contribute a photon to the growing packet of energy. Indeed, the train of waves may continue to accumulate energy by stimulated emission until it travels out of the gas. The argon laser is merely an apparatus designed to encourage the continued growth of the train by causing it to travel back and forth through the gas many times. This effect is achieved by the flat windows and the pair of dielectric mirrors associated with the laser tube. An occasional train of coherent light waves may travel along the axis of the tube and make its way through one of the windows and thence to the adjacent mirror. If the mirror is of high optical quality, it reflects most of the light directly back through the window and into the tube, where the light accumulates still more energy by the process of stimulated emission. The intensified light proceeds through the opposite window, and the cycle of events is repeated.

"Not all the energy of the beam is reflected by the mirrors. Mirrors of perfect reflectivity cannot be made. Some of the energy is absorbed by the reflecting material and transformed into heat. Another portion, perhaps a few thousandths of 1 percent, makes its way through the reflecting material. This small portion constitutes the output of the laser.

"To create the current of 15 to 20 amperes that excites the gas a potential of some 2,000 volts must be applied to the electrodes of the tube. The resulting expenditure of power amounts to several kilowatts—enough to heat the tube beyond the melting point of the glass. In order to prevent destructive heating, power is applied to the laser in short, rather widely spaced pulses. In the design that I recommend the tube is so energized 120 times per second. The pulses persist only a few millionths of a second. Pulses of coherent light are emitted at the same rate and persist for less than 50 millionths of a second. The beam appears continuous to the eye, however, because the relatively sluggish chemical processes of vision cause each pulse to be seen for about a fiftieth of a second.

"The laser consists of a base assembly that supports a capillary tube 50 centimeters long with a bore of two millimeters. Each end of the capillary tube is sealed to a 15-millimeter tube that includes a quartz window and a neon-sign electrode. A ball-and-socket joint of glass in the 15-millimeter tube permits adjustment of the angle the windows make with the axis of the capillary. Air is pumped from the assembly and argon gas is admitted to it through a short length of seven-millimeter tubing sealed into the 15-millimeter tube at one end [*see bottom illustration on opposite page*]. All parts are made of borosilicate glass except the quartz windows.

"Begin the construction by cutting a

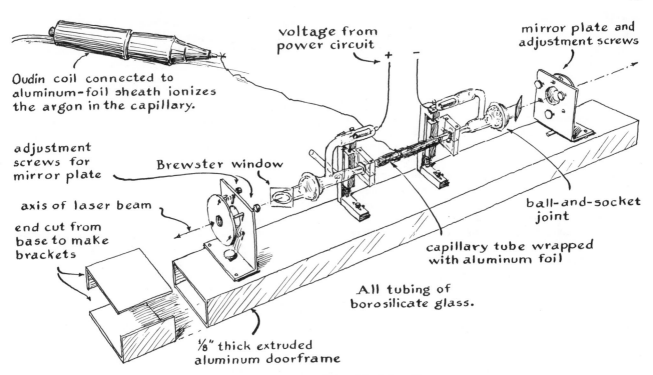

voltage from power circuit

mirror plate and adjustment screws

Oudin coil connected to aluminum-foil sheath ionizes the argon in the capillary.

adjustment screws for mirror plate

Brewster window

axis of laser beam

end cut from base to make brackets

ball-and-socket joint

capillary tube wrapped with aluminum foil

All tubing of borosilicate glass.

⅛" thick extruded aluminum doorframe

*Overall view of the laser designed by Sylvan Heumann*

51-centimeter length of capillary tubing. This tubing comes in standard lengths of four feet. With a corner of a flat file make a crosswise nick in the glass at the specified length, grasp the tubing on each side of the nick and pull it apart. Do not bend the glass. With a blowtorch that burns a mixture of oxygen and household gas, heat the cut ends just enough to round the sharp edges. When the glass cools, reheat one end until the bore closes. Blow into the opposite end to form a bulb about 18 millimeters in diameter. Let the bulb cool until it solidifies. Reheat the outer hemisphere of the bulb to softness, then blow forcefully to explode the softened glass. Strike off with the flat face of the file any tissue-thin fragments that cling to the expanded end of the

capillary. Rotate the expanded end of the tube in the fire until the edge shrinks to a diameter of 15 millimeters. Similarly expand the other end of the capillary.

"Next, select a cork that fits 15-millimeter tubing and bore an axial hole through it to fit a pencil-sized length of wooden dowel rod. Insert the end of the rod completely through the cork from the top. Insert the small end of the cork into the ball end of the ball member of the ball-and-socket joint. Align the dowel with the axis of the ball member so that when the dowel is rotated between the thumb and fingers, the glass turns without wobbling, as though it were in a lathe. Grasp the dowel in one hand and the capillary in the other. Bring the 15-

millimeter tubing of the ball member into axial alignment with the capillary. While rotating the glasses back and forth synchronously, move the ends of both pieces into the edge of the flame on opposite sides and heat the glass until about one millimeter of each edge softens. Remove the glasses from the fire and, while maintaining the back-and-forth rotation, press the aligned ends lightly together so that they fuse. Return the fused joint to the fire. Continue rotating the glass back and forth. Never let it stop. When it becomes soft, remove the work from the fire, stretch the joint about five millimeters and blow into the open end of the capillary until the glass expands into a rounded contour. You have

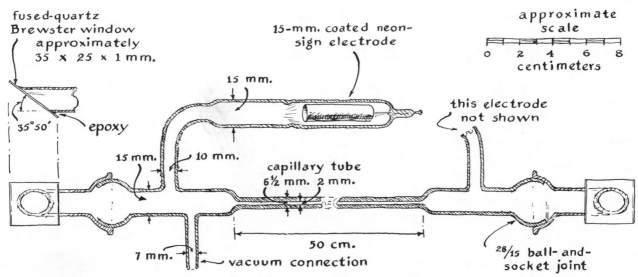

fused-quartz Brewster window approximately 35 x 25 x 1 mm.

15-mm. coated neon-sign electrode

approximate scale

0  2  4  6  8
centimeters

35°50'      epoxy

15 mm.

15 mm.      10 mm.

this electrode not shown

capillary tube
6½ mm.  2 mm.

50 cm.

7 mm.      vacuum connection

²⁸⁄₁₅ ball- and-socket joint

*Details of the laser tube*

now made a 'butt seal.' In the same way seal the remaining ball member to the opposite end of the capillary.

"By means of the same technique seal a 25-centimeter length of 10-millimeter tubing to a 25-centimeter length of 15-millimeter tubing. At a point three centimeters from the seal, cut the 15-millimeter tubing and seal to its end a 15-millimeter neon-sign electrode of the type that is coated internally with a mixture of barium carbonate and strontium carbonate. Similarly prepare a second electrode. Bend the 10-millimeter tubing of each electrode assembly to a right angle by softening a three-centimeter zone of the glass adjacent to the seal. Bend the softened glass by turning the ends upward. (Let gravity work with you.) When the bend is completed, promptly blow into the open end of the tubing to restore the original diameter of the curved portion. When the glass cools, cut the 10-millimeter tubing to the illustrated proportions.

"Seal the electrode subassemblies into the ball members of the capillary assembly. To make this seal blow a hole in the 15-millimeter tubing of the ball member by placing a stopper in the opening of the ball and heating a spot about 12 millimeters in diameter in the middle of the 15-millimeter tubing. Blow the softened spot into a hemisphere. Reheat about 70 percent of the hemisphere and blow forcefully to explode the upper part of the bulb. Butt-seal an electrode subassembly to the hole. Transfer the stopper to the opposite end and seal the second electrode subassembly. Finally, with the same procedure, seal a four-inch length of seven-millimeter tubing into the assembly at the position indicated by the drawing. (The basic techniques of making glass apparatus by hand are fully explained in *Creative Glass Blowing*, by James E. Hammesfahr and Clair L. Stong, W. H. Freeman Company, 1968.)

"The tube is completed by cementing the quartz windows to the socket members of the ball-and-socket joints. The windows transmit the laser beam with substantially no loss of light only if they are cemented to the glass tube at an angle of approximately 35 degrees 50 minutes with respect to the axis of the tube. My tubing was cut to this angle by a diamond saw of the type used by amateur mineralogists. The cuts can also be made with a blade of soft metal, such as brass, that is fed a slurry of No. 120 grit Carborundum and water. The blade, a strip of brass about .02 inch thick and 10 inches long, can be mounted for use in a hacksaw frame. The angle of the cut can be maintained with an improvised miter box. The cut end of the glass must be lapped smooth and flat by grinding the tubing against a sheet of plate glass with a slurry of No. 600 grit Carborundum.

"To cement the quartz windows in place, coat the mating surfaces of the balls and sockets lightly with high-vacuum stopcock grease, connect the laser tube to the vacuum pump through the seven-millimeter inlet and start the pump. Place the previously cleaned windows in contact with the cut ends of the socket members and assemble the sockets simultaneously to the balls. Suction will hold the joints and the windows in place. If this two-handed manipulation proves difficult, have an assistant position one of the windows while you position the other one. If the cut ends have been lapped flat, the tube assembly will be airtight. With a toothpick gently apply a bead of epoxy cement completely around the junction of the tube and window. The cement should not flow into the junction and it will not if the end of the tube has been lapped flat. Let the vacuum pump run until the cement solidifies. This operation completes the laser tube.

"The mechanical assembly can be improvised from almost any available materials. For the base I used a scrap of rectangular aluminum tubing four inches wide, two inches thick and 36 inches long of the type found in metal doors. The laser tube is attached to the base with a pair of adjustable fixtures that are convenient for aligning the tube coaxially with the mirrors. The fixtures can be made with ordinary hand tools. The mirrors are mounted in adjustable cells that include micrometer screws spaced radially at 120 degrees. The optical axis of the mirrors can be positioned as desired by manipulating the screws. Several of the accompanying illustrations give details of the supporting fixtures and mirror cells.

"It is possible to prepare the laser for operation with a fairly crude vacuum system. A mechanical pump capable of exhausting the tube to a pressure of .01 torr is adequate. (One torr is equal to the pressure exerted by a column of mercury one millimeter high.) The useful life of the tube tends to increase, however, with the effectiveness of the vacuum sys-

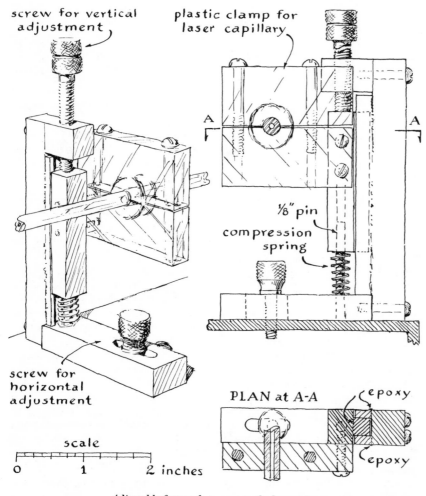

screw for vertical adjustment

plastic clamp for laser capillary

A                    A

⅛" pin

compression spring

screw for horizontal adjustment

PLAN at A-A

epoxy

epoxy

scale

0        1        2 inches

*Adjustable fixture that supports the laser tube*

tem. My laser operates for about 30 minutes on a charge of gas. I rarely disconnect it from the vacuum system. When laser action begins to fail, I replace the argon by opening a stopcock that lets the used gas flow into the pumps. Then I close the stopcock and open another one that admits fresh argon from the reservoir. The apparatus now operates for another 30 minutes.

"My vacuum system includes both a mechanical pump and a diffusion pump, a cold trap, a closed-end manometer, a vacuum gauge, a flask of argon and five stopcocks [see illustration on page 23]. (The diffusion pump, cold trap and gauges may be omitted.) The stopcock that connects the gas reservoir to the system should be of the high-vacuum type and need not have a bore larger than two millimeters. Remove the plug of this stopcock. With a file make a scratch about a third of the way around the plug beginning at the hole. Turn the plug clockwise and, by reducing the pressure on the file, let the depth of the scratch taper gradually to the surface. Make a similar scratch that extends in the same direction from the other end of the hole. Apply a thin film of high-vacuum stopcock grease to the plug and replace it. When the stopcock is operated, gas flows through the scratches slowly, enabling you to fill the tube with argon at a precisely controlled rate.

"Current for exciting the laser tube is drawn from a power supply at the rate of 120 pulses per second. The power supply consists of a variable transformer that feeds a neon-sign transformer. The high-voltage output of the neon-sign transformer is converted to direct current for charging a capacitor. Pulses of current are drawn from the capacitor by the tube when the argon gas is ionized by an Oudin coil, which generates a potential of 30,000 volts at a very high frequency. The amount of current depends on the adjustment of the variable transformer. Any neon-sign transformer can be used that is rated at an output potential of between 4,000 and 9,000 volts and a current of not more than 50 milliamperes. The variable transformer must be rated at a current of at least two amperes. The capacitor may be rated at one microfarad and at a breakdown voltage at least equal to that of the neon-sign transformer.

"A higher potential than is provided by the power supply is required to ionize argon gas; it can be developed from an Oudin coil of the type used for detecting leaks in the glass parts of vacuum systems. An example is the No. 15–340–75V3 Vacuum Tester distributed by the

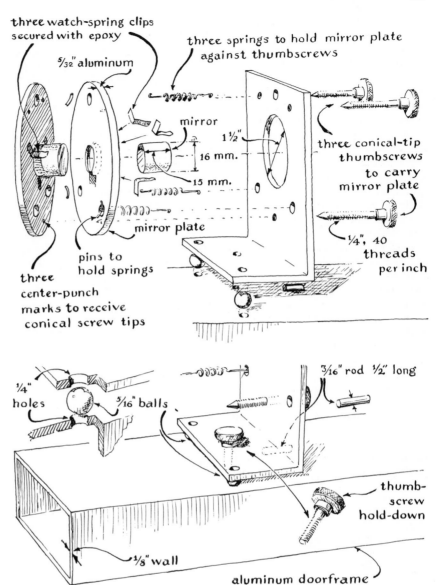

Details of the mirror cell

Fisher Scientific Company, Springfield, N.J. 07081. Connect a wire from the high-voltage terminal of the coil to a piece of aluminum foil wrapped around the middle of the capillary tube. The high potential of the Oudin coil triggers pulses of pumping current from the capacitor. Place the capacitor within a few inches of the tube to minimize the length of the leads that connect it to the electrodes.

"When the laser has been assembled, connect the tube to the vacuum system and install the dielectric mirrors in their cells. One mirror should be spherical and have a radius of 120 centimeters. The other mirror should be flat. Ordinary silvered or aluminized mirrors will not work. With the mirrors installed, place a small incandescent lamp near the end of the tube so that the light that passes through the glass is reflected into the

bore of the capillary by the inclined inner face of the quartz window. Look into the bore of the capillary through the mirror nearest the lamp and adjust the micrometer screws that control the distant mirror until the circles of light that are reflected from the walls of the tube and the bright spot from the distant mirror are concentric. Transfer the lamp to the distant end of the tube and adjust the mirror there.

"Close the stopcock of the gas reservoir and the stopcock leading to the air inlet. Open all stopcocks between the vacuum pump and the tube. Start the pump. With a low-temperature flame of the kind delivered by a propane gas torch heat all parts of the tube except the ball-and-socket joints and the quartz windows. The glass should be made hot enough to turn a piece of white tissue yellow in about 20 seconds. The heat drives adhering gases from the inner walls of the glass.

"The electrodes must now be brought to a dark red heat. I heat them one at a time by sliding over the glass envelope of the electrode a coil of wire that is connected to the output of a 75-watt amateur-radio transmitter. A variable capacitor is connected across the coil for tuning it to resonance with the frequency of the transmitter. The size of the variable capacitor and of the coil depends on the frequency at which the transmitter operates, but a coil of nine turns that is an inch wide and 1½ inches long will work with most transmitters when it is tuned by a variable capacitor of about 150 picofarads.

"The electrode will begin to heat as the circuit approaches resonance. At full resonance the electrode may become hot enough to melt, so tune the circuit cautiously. Stop tuning at the point where the electrode becomes red hot. The heated electrode will liberate an astonishing amount of gas, enough to alter the characteristic sound of the vacuum pump. Maintain the metal at red heat for about three minutes, then similarly 'outgas' the second electrode. In addition to liberating gas from the metal, the heat burns away the nitrocellulose binder used for coating the interior of the electrode and converts some of the barium carbonate to metallic barium. Part of the metallic barium then combines with chemically active gases and vapor, such as oxygen, nitrogen, carbon dioxide and water vapor, that remain in the tube.

"To fill the tube with argon turn on the power supply and the Oudin coil, adjust the variable transformer for an output of approximately 70 volts and close the stopcock to the vacuum pump and the one to the trap. Open the stopcock to the argon flask just enough to allow a flow of gas. As argon enters the system the tube will begin to glow faintly and will increase in brightness as the gas pressure increases. The rate of gas flow differs with various stopcocks and must be determined experimentally.

"The tube may flash intermittently and the gas may change from light blue to a pinkish hue. The color change indicates the presence of an unwanted gas, such as oxygen. If the color change occurs, shut off the gas supply, let the tube operate for about a minute and then shut off the power. Open the stopcock to the vacuum pump and exhaust the system for about five minutes. Repeat the sequence of operations until the tube glows dark blue and retains this color for at least five minutes. Pump out the gas.

"The apparatus is now ready for laser action. With the variable transformer turned on and adjusted for an output of about 70 volts, admit argon gas to the tube as slowly as possible. The tube may glow dimly and flicker. Continue to add argon just to the point where the tube stops flickering. If the mirrors have been adjusted perfectly, a beam of greenish-blue coherent light will be emitted at the ends of the tube.

"If the beam does not appear, twist the micrometer screws (one at a time) back and forth about two or three angular degrees. This operation is known as 'fiddling' with the screws. Watch the ends of the tube carefully as you rock the screws. Stop when the beam appears. Then adjust the system for maximum beam intensity. Increase the output voltage of the variable transformer. Beyond a certain maximum voltage the intensity of the beam will decrease. Then, assuming that you have installed the quartz windows so that the faces are at right angles to the vertical plane, you will observe two spots of light on the ceiling. Rotate the windows so that the spots lie in the vertical plane that includes the bore of the capillary. Then tilt each window up or down to the angle at which the beam becomes most intense. Both of these adjustments are made possible by the ball-and-socket joints.

"Next, vary the gas pressure. Add gas slowly. Up to a certain pressure the intensity of the beam will increase. Thereafter it will decrease. Lower the pressure by pumping out gas. After 20 or 30 minutes of operation at maximum intensity the brightness of the beam may begin to decline. This indicates that atoms of argon are being buried under particles of eroded electrode materials, an effect known as 'sputtering.' The sputtering effect lowers the pressure of the gas. At this point I usually pump out the tube and refill it with fresh argon.

"You will observe that the color of the beam changes as the input voltage is increased. The change is caused by the successive appearance of coherent light at various wavelengths as current in the tube is increased. Laser action begins when the current reaches 1.45 amperes per pulse. The tube then emits coherent light at a wavelength of 4,880 angstroms, the wavelength for which the dielectric mirrors are coated. At 3.6 amperes laser action also begins at 5,145 angstroms. Thereafter light appears at the following wavelengths: 3.8 amperes, 7,465 angstroms; 4, 4,965; 5.2, 4,579; 6, 5,017; 6.9, 4,658; 10.5, 5,287, and 15 amperes, 4,727 angstroms. Each color can be separated into an individual beam by passing the output of the laser through a 60-de-

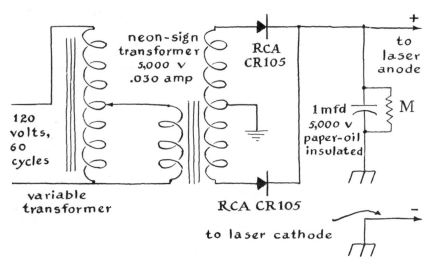

*Circuitry of the power supply*

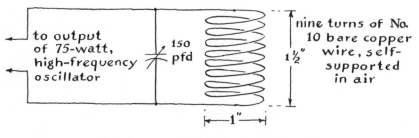

*Radio-frequency circuit for heating the electrodes*

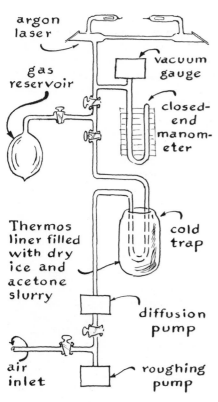

argon
laser

gas
reservoir

vacuum
gauge

closed-
end
manom-
eter

cold
trap

Thermos
liner filled
with dry
ice and
acetone
slurry

diffusion
pump

air
inlet

roughing
pump

*The vacuum system*

gree glass prism. The individual beams can be used for the precise measurement of length. Collectively they will serve as a series of known wavelengths for calibrating apparatus and making numerous other experiments. Krypton gas can be used in this laser and will demonstrate still other spectral lines, but with lesser beam intensity.

"The dielectric mirrors, ball-and-socket joints, electrodes sealed in Pyrex, transformers, gas and other supplies can be bought from Henry Prescott, 116 Main Street, Northfield, Mass. 01360. A note of warning: The laser beam is hazardous. It can burn and destroy the retina of the eye. Two intense beams are emitted from the ends of the tube and two beams of lesser intensity are reflected into the room by the faces of the quartz windows. Mask all beams close to the tube except those with which you are experimenting. Avoid spurious reflections of the experimental beam by objects in the room. The power supply develops lethal voltages and the capacitor stores charge at high voltage. Handle them accordingly. When the apparatus is turned off, always short-circuit and thus discharge the capacitor by means of a wire supported in an insulating handle. As an added precaution it is well to connect a one-megohm, two-watt resistor permanently across the terminals of the capacitor to 'bleed off' automatically any charge that may accumulate spontaneously."

# 4

# Tunable Dye Laser

*An inexpensive tunable laser made
at home using organic dye*

*February 1970*

An ideal laser, from the amateur's point of view, would be easy to make, reasonably inexpensive and capable of generating light of any color. An instrument of this kind has been built by J. R. Lankard of the International Business Machines Corporation. He made it at home with ordinary hand tools for less than $75, largely to prove that it could be done.

The amplifying medium of the laser is a dilute solution of organic dye that is periodically "pumped," or illuminated, by a homemade flash lamp. The output is an intense beam of coherent light five millimeters in diameter that can be focused to a point with lenses. The sharpness of the point is limited only by the wave nature of light. The beam can also be spread into a pattern of diverging rays. The color of the coherent rays is determined by the nature of the dye.

Various dyes can be used. Each of them generates light that spans a characteristic portion of the spectrum. The laser can be tuned to generate light of any wavelength within these ranges by means of a diffraction grating. When the laser is pumped at a sufficiently high rate, say 60 pulses per second, the output beam appears to the eye to be continuous. This modification increases the cost substantially, because the size of the power supply must be increased and provisions must be made for cooling the unit.

Like the helium-neon laser, the dye laser consists essentially of a slender amplifying tube, or dye cell. The ends of the tube are closed by a pair of flat windows that face a pair of mirrors. A second tube, the flash lamp, is mounted parallel to the amplifying tube. Both tubes are surrounded by an aluminum tube of elliptical cross section with an interior surface that is highly polished to function as a mirror. The axis of the amplifier tube occupies one focus of the elliptical mirror and the axis of the lamp occupies the other focus. White light emitted by the lamp is thus concentrated by the mirror on the amplifying tube [*see illustrations on opposite page*].

The amplifying tube is filled with an alcohol solution of dye that is known to fluoresce. When such solutions absorb white light, some molecules of dye acquire an abnormal amount of energy and so are raised to what is termed their lowest excited singlet state. Later they spontaneously emit the excess energy as light of longer wavelength. This is the phenomenon known as fluorescence.

If the white light is sufficiently intense, as it is in the dye laser, a majority of the dye molecules absorb more than enough energy to reach the singlet state. Hence they are raised to one or another of the possible higher energy states. From these levels they can also drop back spontaneously to lower energy states by emitting light. Some of these emitted waves, strictly by chance, head toward one or the other of the laser's flat mirrors, which reflect the light waves back into the amplifier tube. Here the waves stimulate other excited molecules to emit excess energy. The resulting emission is identical in color, and so in wavelength, with the stimulating waves. The waves merge and thereafter proceed in lockstep. In effect, the waves that were emitted spontaneously are amplified and continue to accumulate energy as they are subsequently reflected back and forth through the tube [see "Organic Lasers," by Peter Sorokin; SCIENTIFIC AMERICAN, February, 1969].

The flat mirrors of the dye laser are coated with aluminum. One mirror reflects about 92 percent of the incident light and absorbs 8 percent. The other mirror, which has a thinner coating, reflects 74 percent, absorbs 8 percent and transmits about 18 percent. The transmitted portion constitutes the output beam of the laser. In Lankard's apparatus it amounts to about 5,000 watts per pulse, which is a million times more intense than the output of the helium-neon laser! This substantial power and the tunable feature of the laser open new experimental opportunities to amateurs who have been restricted to the helium-neon and argon lasers previously described in this department [see "The Amateur Scientist"; SCIENTIFIC AMERICAN, September, 1964, December, 1965, February, 1967, and February, 1969].

It bears repeating that laser action occurs only if the dye is exposed to white light of sufficient intensity. Lankard developed the required intensity with an extraordinarily simple lamp. It is a tube of fused quartz provided with stainless-steel electrodes and filled with air. The electrodes are connected to the terminals of a 15-microfarad capacitor charged to a potential of 3,000 volts. To flash the lamp Lankard pumps air from the tube. When the pressure falls to about 60 torr, the capacitor discharges through the lamp. The resulting flash consumes some 35 megawatts of power and persists for about two millionths of a second. The capacitor is specially designed to discharge in that brief interval. Ordinary capacitors discharge more slowly and will not work in this application.

The dye emits only a small part of the absorbed energy as coherent light. That part is transformed into heat that causes random differences in density throughout the column of dye and, as a result, random differences in the optical refraction of the fluid. Fortunately the heating effect lags behind the flash by several millionths of a second. By the time the

optical distortion becomes substantial
the laser light has been emitted.

The dye must be at uniform temperature, however, before the laser can be
pulsed again. A convenient method of
cooling is to circulate dye continuously
through the amplifier tube. The tube
consists of an 80-millimeter length of
seven-millimeter quartz tubing with a
bore of five millimeters. The ends of the
tube are closed by flat windows cemented in place at the edges with epoxy cement. Dye solution flows into one end
of the tube through a port in the side and
out through a similar port at the other
end.

The ports can be made by welding
short lengths of quartz tubing, as side
arms, near the ends of the amplifier tube.
Fused quartz is difficult to soften and
manipulate. It must be worked in an
oxyacetylene flame. Lankard suggests
the use of metal tubing as an alternative.
A T fitting can be improvised from copper or stainless-steel tubing. The inside
diameter of the T should be seven millimeters to match the outer diameter of
the quartz tube. One end of the crossbar
of the T can be slipped over the end of
the quartz tube and fixed in place with
epoxy. The quartz window, which has a
diameter of about 10 millimeters and a
thickness of one millimeter, is put on the
remaining end of the crossbar and held
in place while a coat of epoxy is applied
to the joint. A short nipple of copper or
stainless steel forms the leg of the T and
functions as the port through which the
dye solution flows [see top illustration
on next page].

Dye solution can be circulated
through the laser by a small pump. The
cost of the pump can be avoided by si-
phoning dye solution from an elevated
container and discharging it into a similar container at lower elevation. The
dye should flow at the rate of about one
liter in 15 minutes.

The pump that evacuates the lamp
tube can be the sealed compressor from
a discarded electric refrigerator that is in
reasonably good condition. Connect the
exhaust port of the lamp to the inlet tube
of the compressor. Usually the inlet tube
is the larger of two copper tubes welded
into the sealed unit as a pair. The smaller
of the tubes resembles a wire about 3/32
inch in diameter; it is actually a capillary that serves as the outlet of the compressor. Cut the capillary by filing a nick
about 1/32 inch deep and bending the
tube at the nick until it breaks. Run a
small hose from the capillary to a bucket
of soapy water.

Oil of any kind—indeed, any contamination of the dye—will suppress laser ac-

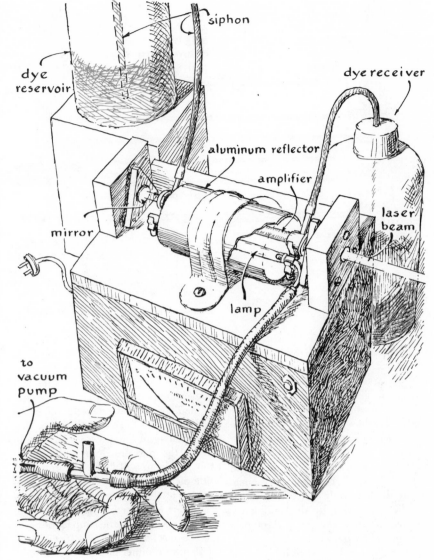

*Elements of the dye laser made by J. R. Lankard*

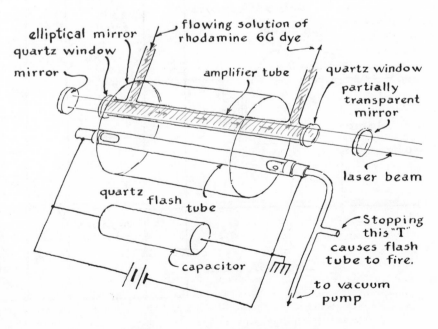

*Arrangement of the laser's components*

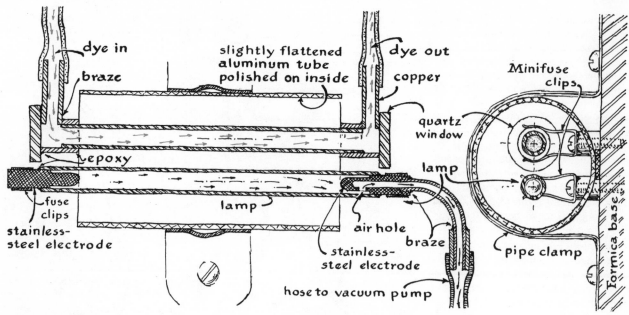

*Details of the trigger lamp and the amplifier*

tion. Do not use ordinary plastic tubing in any portion of the dye plumbing. Most plastics contain an oily plasticizer that preserves the flexibility of the material. Use glass reservoirs and connect them to the amplifier with either glass or Kodak Polyflow tubing, a plastic that contains no plasticizer. Do not attempt to operate the laser in a room where the air may be contaminated with oil, and remember that mechanical vacuum pumps discharge oily fumes. Such fumes can be minimized by discharging exhausted air from the pump through a container of soapy water.

The lamp can be flashed by either of

two triggering schemes. In the simpler and cheaper of the two a T fitting is installed in the tube that leads from the lamp to the air pump. Cut the tube at a convenient point and insert the crossarm of the T. Air will be pumped through the open leg of the T. To flash the lamp, plug the opening of the T with your thumb. When the pressure drops to about 60 torr, the lamp will flash. The grounded side of the power supply must be connected to the electrode through which the air is exhausted or you will get an electric shock when the lamp flashes.

The second scheme is a special triggering transformer in the power supply.

This device develops high potential, much like a spark coil, when power is applied to its primary winding. One terminal of the transformer is connected to ground and the other one to a few turns of fine wire wrapped around the middle of the lamp. The lamp is exhausted to a pressure about 10 torr above its normal firing pressure, which must be determined experimentally. Then, with full voltage applied across its terminals by the charged capacitor, the lamp can be fired by applying power to the primary winding of the triggering transformer. High voltage ionizes air inside the lamp and initiates the discharge, as indicated

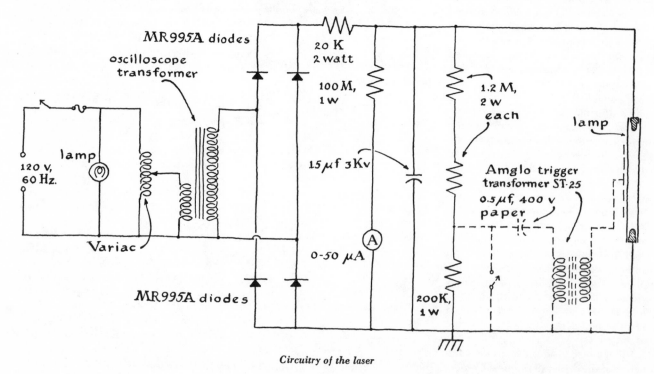

*Circuitry of the laser*

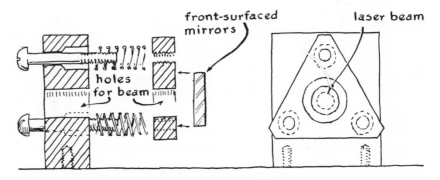

*Construction of the mirror cells*

by the broken lines in the illustration below. The scheme is convenient when an experiment requires that the laser be fired electronically by an associated apparatus that includes a switch for closing the triggering circuit.

The lamp must have a port for exhausting the air. A side arm of quartz can be welded to the tube. Alternatively, the air can be withdrawn through one of the electrodes. The electrodes are preferably made from small rods of stainless steel that make a snug fit with the bore of the quartz tube. The diameter of the inner portion of both electrodes should be reduced to about four millimeters through a length of about eight millimeters and the end rounded into a polished hemisphere. The portion of reduced diameter, along with about six millimeters of the full diameter of the rod, is inserted into the end of the quartz tube and sealed in place with an external coat of epoxy cement. The reduced diameter provides a 1/2-millimeter space between the metal and the quartz. The space prevents hot plasma from concentrating in the zone where the metal and the quartz make contact. The portion of the rod that extends outside the tube may be of any length convenient for the attaching of leads from the capacitor.

The machining can be done by clamping the rod in the chuck of an electric drill and cutting the metal with a file as the drill turns. Air can be exhausted through a hole about a millimeter in diameter drilled partway through the axis of the rod and joined at a right angle by a hole of similar size drilled from a point close to the shoulder. Do not drill the axial hole completely through the electrode. A tube of copper capillary can be brazed to the outer end of the electrode for attaching the air pump. Connect this electrode to the grounded side of the power supply.

The lamp and the amplifying tube are mounted parallel, 15 millimeters above the base, with their centers separated by about 12 millimeters. The base can be a slab of Formica about 1/2 inch thick. Lankard clamps the tubes in Minifuse clips attached to the base.

The reflector consists of an 80-millimeter length of aluminum tubing that has an inner diameter of about 25 millimeters. The inner surface must be highly polished; the polishing can be done with a small buffing wheel. Bring the surface to a semipolish by applying tripoli to the buffing wheel. After washing the metal with soap and water complete the polish with a clean buff to which rouge is applied.

The polished tube must be converted into an elliptical mirror. Clamp it in a vise and by trial and error exert just enough pressure to deform the metal. When the pressure is released, the tube should spring into an ellipse with its major axis some three millimeters longer than its minor axis. If you squeeze it too much, so that the difference in length is more than three millimeters, correct the error by rotating the tube a quarter of a turn in the vise and squeezing it again.

Mount the reflector to the Formica base with a pipe clamp. The major axis should be made parallel to the base. The Minifuse clips are attached to the base at points such that the axes of the amplifier and the lamp coincide with the focuses of the ellipse, which lie 12 millimeters apart.

Cells for supporting and adjusting the position of the mirrors consist of two aluminum plates fastened together with machine screws and helical springs of the compression type [*see illustration above*]. One of the plates, a rectangle that serves as a pillar, is fastened to the Formica base with machine screws. Three holes spaced 120 angular degrees apart admit machine screws that engage threads in the second plate, which can be of triangular form. Compression springs that surround the screws hold the plates apart. A hole about eight millimeters in diameter is drilled through both plates of one cell. The hole is centered in the triangular plate. The partially transmitting mirror is cemented at its edges to this triangle, with the aluminum coating facing the dye cell. The hole functions as an aperture for the output beam. The fully transmitting mirror is similarly cemented with epoxy to the remaining cell.

The power supply can be assembled in a metal box that supports the Formica base. It is extremely important to use the shortest possible leads for connecting the 15-microfarad capacitor to the lamp, preferably leads of copper strap about one millimeter thick and 10 millimeters wide. The leads can be cut from copper flashing of the kind available from lumberyards and tinsmiths. Attach the leads to the screws that fasten the Minifuse clips of the lamp to the base. This construction minimizes the electrical inductance of the circuit and the interval required for the capacitor to discharge. The intensity of the light varies inversely with the rate of discharge.

The microammeter, which can be mounted in one side of the box, functions as a voltmeter, the scale indicating hundreds of volts. A reading of 30, for example, indicates that the capacitor is charged to a potential of 3,000 volts. A potential of 120 volts applied to the primary winding of the oscilloscope transformer appears as 2,400 volts across the terminals of the secondary winding. The rectified voltage that appears across the capacitor is equal to the output potential of the oscilloscope transformer divided by .707; a potential of 120 volts applied to the primary winding develops 2,400/.707, or 3,400 volts, across the capacitor.

This potential exceeds the breakdown rating of the capacitor. To protect the capacitor, Lankard inserts a variable transformer between the power line and the power supply and adjusts the potential to 3,000 volts as the capacitor accumulates charge. Do not omit the three resistors shown at the right in the bottom illustration on the opposite page. They function as bleeders, draining charge slowly from the capacitor. Without this safety provision the capacitor would retain a substantial portion of its lethal charge indefinitely. Never touch the power-supply circuit until the capacitor has been short-circuited, even though the circuit includes bleeder resistors. Short-circuiting can be done by briefly connecting a short length of copper wire across the terminals of the capacitor. The wire should be fixed to a dry wooden handle about a foot long.

Several dyes have been used successfully in the laser, and others are under investigation. A good one for a beginning experiment is rhodamine 6G. This orange dye emits light that spans a spectral range of some 440 angstroms from yellowish-green to red. The molecular weight of the dye is 449. It is used as a $10^{-4}$ molar concentration in methanol. The concentration can be achieved by dissolving .045 gram of the dye in methanol to make one liter.

An interesting dye, which emits a strong blue laser beam that is tunable from 4,300 to 4,900 angstroms, is 7-diethylamino-4-methylcoumarin, a dye found in commercial detergent whiteners. It has a molecular weight of 231.3 and is used in the laser at a concentration of 75 milligrams of dye per liter of methanol. Another dye, sodium fluorescein, is used at the same concentration as rhodamine 6G but in an ethanol solution. The molecular weight of sodium fluorescein is 370. Laser emission from this dye centers on 5,500 angstroms. Mix the dyes in clean glass containers, and be careful to avoid oily contamination.

The operation of the fully assembled laser requires one critical adjustment: the mirrors must be positioned so that their surfaces are exactly parallel and perpendicular to the axis of the dye cell. When the surfaces are so aligned, light rays that are emitted parallel to the axis of the amplifier tube are reflected back on themselves and thereafter oscillate between the reflecting surfaces as though trapped in a cavity. Several schemes have been devised for aligning the mirrors. The most convenient one involves the use of an instrument consisting of a small telescope, an optical beam splitter and a source of light. Lankard's instrument consists of a seven-power telescope that is available from the Edmund Scientific Co., 600 Edscorp Building, Barrington, N.J. 08007 (catalogue number 50,249). The telescope comes with a diagonal mirror that deflects the incoming light at a right angle into the eyepiece.

Lankard substituted for the diagonal mirror a beam splitter that is also available from Edmund Scientific, and behind the beam splitter he installed a pinhole aperture and a small incandescent lamp [*see illustration below*]. Rays from the lamp proceed through the pinhole, the beam splitter and the objective lens of the telescope. When the telescope is aimed at a distant mirror that reflects the light back into the telescope, a portion of the incoming light is diverted into the eyepiece by the beam splitter, and an image of the pinhole appears in the eyepiece.

To align the laser, Lankard removes the partially transmitting mirror and, with the telescope, looks through the axis of the empty dye cell at the distant fully reflecting mirror. That mirror is then adjusted to center the image of the pinhole in the eyepiece. The partially transmitting mirror is reinstalled. Two images of the pinhole, usually displaced, now appear in the eyepiece. The partially transmitting mirror is adjusted to bring the two images into exact register. This completes the adjustment. When the laser is thus adjusted, it will emit a pulse of coherent light each time the lamp flashes. The beam appears yellowish in the case of rhodamine 6G dye.

The laser can be tuned to emit any desired spectral line within the range of the dye by substituting a reflecting diffraction grating for the fully reflecting mirror. The grating, which need be no larger than 10 millimeters square, is mounted so that it can be rotated in three planes: its own plane, the horizontal plane and the vertical plane [*see illustration at right*]. The rulings are placed exactly parallel to the horizontal axis of rotation. (The grating, which must be obtained commercially, should be ruled

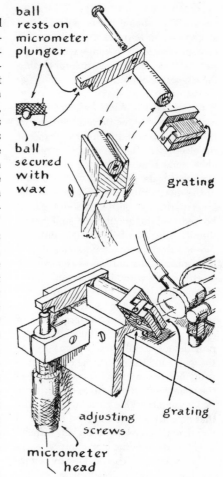

*Mounting of the diffraction grating*

with at least 1,800 lines per millimeter and blazed for 5,000 angstroms in the first order.) When the grating is properly aligned, it reflects light of a single spectral line into the amplifier. The color depends on the horizontal angle the grating makes with the axis of the amplifier tube.

Initially the grating can be set to an angle of about 70 degrees. A colored image of the filament will appear in the alignment telescope. When the image is brought into register with the white image reflected by the partially transmitting mirror, the laser will emit light of the selected color. The output beam can then be tuned to any other part of the available spectrum (from a shade of green through yellow and to red) by rotating the grating on its vertical axis.

The laser can also be tuned within broad limits by altering the concentration of the dye or—what amounts to essentially the same thing—by altering the length of the amplifier tube. In general the wavelength of the output beam increases with the concentration of the dye or the length of the amplifier tube.

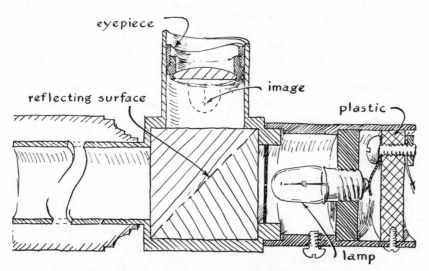

*Arrangement of the telescope for aligning the mirrors*

All the materials required for the construction of the dye laser, with the exception of the telescope, beam splitter and diffraction grating, can be obtained from Henry Prescott, 116 Main Street, Northfield, Mass. 01360. Many parts can be improvised from odds and ends. Lankard's instrument came largely from his scrap box. Do not attempt to economize on the 15-microfarad capacitor. Capacitors characterized by higher inductive reactance will not work. Do not substitute other materials for fused quartz. Finally, avoid the laser beam as if it were the flame of an oxyacetylene torch. It can fry tissue, including the tissue in your eye.

## Note on the Power Circuit

*April 1970*

The circuit diagram of the power supply for the tunable laser using organic dye, described in this department in February, shows a connection between the primary and secondary windings of the oscilloscope transformer. The connection invites the destruction of one of the diodes and should be omitted, as several readers have pointed out.

# 5

# Carbon Dioxide Laser

*A carbon dioxide laser constructed
by a high school student*

September 1971

Numerous amateurs have undertaken the formidable but nonetheless fascinating task of making a gas laser. Earlier articles in this department have described how to build a helium-neon laser and an argon laser [see "The Amateur Scientist"; SCIENTIFIC AMERICAN, September, 1964, and February, 1969]. Now Jeffrey Levatter, a high school student in Encino, Calif., has made a carbon dioxide laser.

The carbon dioxide laser produces a beam not of light but of infrared radiation. It is somewhat easier for the novice to build than the helium-neon laser because it involves no glassblowing. Moreover, it is relatively inexpensive. The costly dielectric mirrors of the helium-neon laser are replaced in the carbon dioxide laser by copper-coated mirrors that can be made at home. Levatter explains the operation of his laser and provides the details of its construction as follows:

"Physically all gas lasers are much alike. A glass tube filled with gas at low pressure is positioned between a pair of facing mirrors. The gas is excited by an electric discharge. Some particles of gas acquire energy by colliding with speeding electrons that are liberated by the discharge. After a finite interval particles thus excited spontaneously emit part or all of the acquired energy in the form of radiation. In so doing a specific particle may either drop to an intermediate level of energy or return to the lowest energy state: the ground level. In effect the gas absorbs energy from the electric circuit and subsequently liberates the energy as radiation in the form of the small packets called photons.

"The energy of the emitted photon depends on the spacing of the energy levels through which the gas particles characteristically fall. A gas particle that is excited to a high energy level can be stimulated to drop to an intermediate energy level if it interacts with a photon of appropriate energy. In dropping to the intermediate level the excited particle emits a photon that is identical with the stimulating photon. The two photons fall into lockstep. The coherent bundle of radiant energy continues to grow by accretion as it encounters still other appropriately excited particles and reacts with them.

"In laser action a growing train of such coherent electromagnetic waves is reflected back and forth through the excited gas by mirrors at the ends of the gas column. Part of the coherent energy escapes through a small window in one of the mirrors. This loss restricts the maximum energy that can accumulate between the mirrors and constitutes the output beam of the laser.

"The efficiency of a gas laser is determined in part by the nature of the gas. In a helium-neon laser the active particles are atoms of neon. To emit infrared radiation neon atoms must be excited to an energy level far above the ground state. Subsequently the atoms emit infrared radiation by dropping a relatively short distance to an intermediate level. The atoms must then return to the ground state before they can again participate in infrared emission. In returning to the ground state from the intermediate level the atoms emit excess energy that makes no direct contribution to the desired infrared radiation.

"In contrast, a molecule of carbon dioxide can be excited to an energy level that lies only a short distance above its ground state. From this level the molecule can emit infrared radiation by dropping a comparatively substantial distance to an intermediate level that lies close to the ground state. For this reason the efficiency of the carbon dioxide laser is impressively greater than that of the helium-neon laser.

"The electric discharge that energizes the system excites carbon dioxide molecules to various energy levels, including the level from which they drop in the course of emitting the desired coherent radiation. Particles excited to the other levels make no direct contribution to the output, although much of that energy is conserved. A significant portion of it is transferred by random collision to previously unexcited molecules, which are thereby raised to the level where they can contribute to the output of the laser. The molecules from which the energy is transferred return to the ground level.

"Although such transfers conserve some of the input energy, efficiency can be improved by mixing other gases with carbon dioxide, notably nitrogen and helium. In effect these gases absorb just the right amount of energy from the electric discharge to raise a carbon dioxide molecule from the ground level to the level whence the molecule can drop by emitting the desired radiation. The transfer of energy occurs during collisions among the several particles.

"As a consequence of the relatively low level to which carbon dioxide must be excited to induce laser action, together with the fact that selective excitation can be achieved by the introduction of other gases, the carbon dioxide laser converts about 20 percent of the input power into coherent radiation. The output power is impressive. My laser develops an infrared beam of about eight watts, which is thousands of times more powerful than the visible output of a helium-neon laser.

"Since the infrared beam is invisible the variety of experiments that can be done with the apparatus is restricted. On the other hand, the high power of the beam invites experimentation of a kind that cannot be achieved with equivalent lasers that operate in the visible part of the spectrum. The beam quickly chars wood. By focusing the rays with an appropriate concave mirror the energy density can be increased to several kilowatts per square centimeter, which is sufficient to burn holes through thin metal. The earth's atmosphere is exceptionally transparent to electromagnetic radiation in the portion of the spectrum extending from a wavelength of eight to 14 microns. Hence the output of the carbon dioxide laser is ideal for communications experiments and also for experiments involving echo ranging.

"It should be possible to make holograms in infrared. Photographic film that is sensitive to infrared radiation is commercially available. I do not know if the grain size of the emulsion is fine enough for adequate resolution, but I look forward to trying the experiment.

"The laser assembly includes a gastight plasma tube in the form of a glass pipe cooled by a water jacket [see illustration below]. The ends of the plasma tube are closed by a pair of metal cells that support the mirrors. Each cell includes a flexible bellows and a set of three screws for adjusting the orientation of the mirrors. The metal cells also serve as electrodes for applying high voltage to the gas. The electrodes are enclosed in boxes of clear plastic to prevent accidental contact with the high potential. The assembly is supported by

an insulating base of wood.

"The borosilicate glass pipe is 18 inches long, with an inside diameter of one inch. It has slightly flared ends, which are sealed to aluminum flanges with silicone-rubber gaskets [see illustration on next page]. The glass, known as Pyrex conical piping, is made by the Corning Glass Works. The central portion of the pipe is surrounded coaxially by a 12-inch water jacket of aluminum tubing two inches in diameter. The ends of the aluminum tube are closed by silicone-rubber caulking. Opposite ends of the water jacket are fitted with pipe nipples of aluminum tubing fastened in place and sealed with epoxy cement. These pipe nipples function as inlet and outlet ports for circulating cold water through the jacket assembly.

"The flange assemblies that clamp to the ends of the glass pipe can be machined out of any metal. Brass is convenient because it solders readily to the copper-coated steel bellows. If aluminum is used for the flanges, the bellows can be sealed in place with epoxy. The ends of the bellows are about an inch in diameter. Bellows of this kind are available from Pathway Bellows, Inc. (P.O. Box 1090, 1452 North Johnson Avenue, El Cajon, Calif. 92020).

"The adjustable flange of the cells supports on its outer face a removable mounting plate to which the mirrors can be sealed with either epoxy or silicone caulking. A circular groove is machined in the outer face of the adjustable flange to accept a rubber O ring that makes a gastight seal between the flange and the mounting plate. Mounting plates are a convenience during the alignment of the optical system because they enable the operator to remove the mirrors.

"The three adjustment screws of each cell are radially spaced at 120 degrees of angle. The threads of the screws engage threads in the adjustable flange. The conical tips of the screws bear against conical indentations in the fixed flange. The screws should have at least 32 threads per inch and should be long enough to place the bellows in tension.

"The laser is fitted with two mirrors, one concave and one flat. The diameter of both mirrors should be somewhat larger than the bore of the plasma tube. The focal length of the concave mirror must be more than twice the distance between the mirrors. Both mirrors can be made of glass coated with a reflective film of either copper or gold. Mirrors coated with gold are available commercially from Esco Products (Oak Ridge

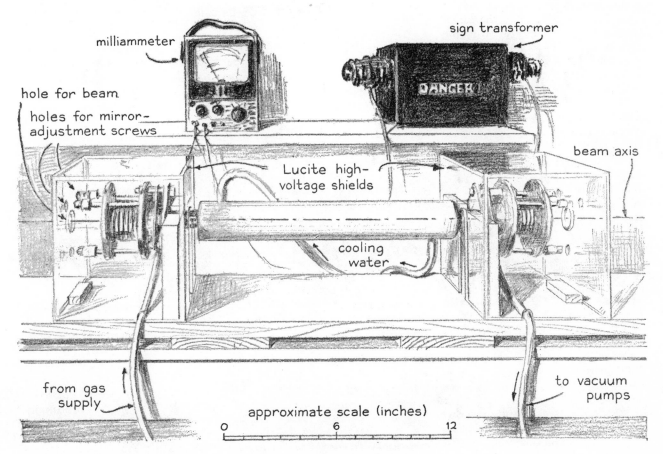

*The laser made by Jeffrey Levatter*

Road, Oak Ridge, N.J. 07438). My experience indicates that copper has higher reflectivity than gold at a wavelength of 10.6 microns.

"The flat mirror transmits the output beam of the laser. It can be made of polished germanium, a material that reflects approximately 60 percent of the incident radiation and transmits 35 percent at 10.6 microns. (The remaining 5 percent is absorbed.) Germanium is expensive. My output mirror consists of a flat disk of polished glass 1/4 inch thick perforated in the center by a hole 3/32 inch in diameter. Glass disks appropriate for this purpose are available from the Edmund Scientific Co. (Barrington, N.J. 08007). The disks are identified as catalogue No. 30,451. The price is 50 cents per disk.

"The flat mirror can be made by drilling a hole through one of the disks. To drill the hole coat the glass with a protective film of pitch or some other waxy material and make a cofferdam around the upper edge with plastic modeling clay. Chuck a short brass rod about 5/64 inch in diameter in a drill press. Fill the cofferdam with a slurry of 220-grade Alundum grit in water. Gently lower the spinning rod into contact with the glass. Raise and lower the rod at one-

second intervals until the abrasive grinds through the disk. Remove the wax with solvent and clean the glass thoroughly. A highly reflective film of copper can be applied to either of the polished surfaces by means of the sputtering technique [see "The Amateur Scientist"; SCIENTIFIC AMERICAN, October, 1967].

"The perforation must be closed on the outer surface of the mirror by a window that is transparent to infrared radiation. Windows made of crystals of sodium chloride or of potassium chloride are effective. Such crystals are hydroscopic, however, and must be kept dry when the laser is not in use. I store my window in a plastic bag that contains anhydrous calcium sulfate as the desiccant.

"Crystals of barium fluoride are much less hydroscopic but absorb substantially more infrared energy. Crystals of appropriate size for making the window are available, both polished and unpolished, from the Harshaw Chemical Company (18051 East Fourth Street, Tustin, Calif. 92680). The price of a large, unpolished crystal of rock salt is about $5. On request the company will send with the crystals an article describing the grinding and polishing of salt

windows. The crystal can be cut at any angle with respect to its axes and need be only large enough to cover and seal the perforation. I suggest that the window be cemented in place with General Electric silicone adhesive, primarily because the cement can be easily removed. This adhesive is usually available in hardware stores.

"The concave mirror can be ground and polished at home by the techniques described in *Amateur Telescope Making: Book One*, edited by Albert G. Ingalls, which is available from *Scientific American*. Two glass disks are required; one serves as a tool for grinding abrasive against the other. After the desired curvature has been achieved and the glass has been ground to a velvety texture by the use of successively finer grains of abrasive the surface of the tool is coated with pitch. The mirror is then polished with rouge applied by the pitch tool.

"After thorough cleaning the polished surface can be coated with copper by the sputtering technique. The copper coating must be thick enough to prevent infrared radiation from penetrating the metal, otherwise the glass may absorb heat and shatter. Time the interval required to deposit a coating that is

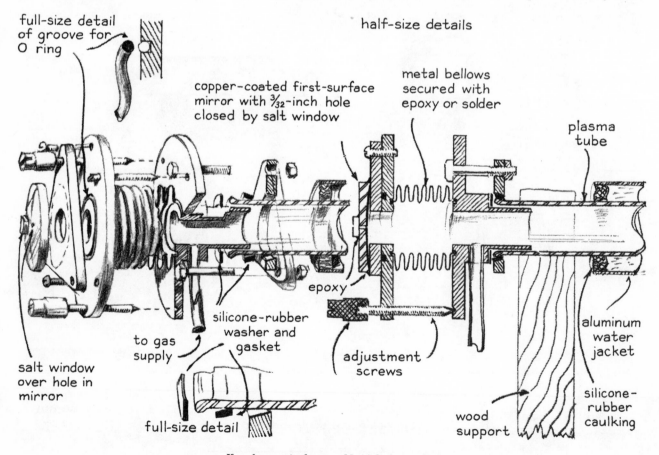

full-size detail of groove for O ring

half-size details

copper-coated first-surface mirror with 3/32-inch hole closed by salt window

metal bellows secured with epoxy or solder

plasma tube

epoxy

salt window over hole in mirror

to gas supply

silicone-rubber washer and gasket

adjustment screws

aluminum water jacket

wood support

silicone-rubber caulking

full-size detail

*Key elements in the assembly of the laser tube*

opaque to visible light and continue to sputter for at least as long.

"A mixture of gases is pumped through the plasma tube continuously when the laser is in operation. Ports for admitting and exhausting the gas are made in the cells. The working pressure of the gas ranges from one to 20 torr, depending on the proportions of the mixture. (One torr is the pressure ex-erted by a column of mercury one milli-meter in height.)

"I draw the gases from three sources. Carbon dioxide sublimes from dry ice in a flask. Nitrogen is obtained from fil-tered air. (Oxygen, water vapor and oth-er gases dilute the nitrogen but do not appear to reduce the output of the laser. Indeed, the power of the beam tends to increase when the incoming air is bub-bled through water. I rarely use water, however, because the vapor can damage the salt window.) Helium is drawn from a cylinder of compressed gas. All gases are admitted through needle valves to a manifold that connects to the laser. Gas pressure is measured by a closed-end manometer filled with vacuum oil and calibrated in millimeters of mercury [see illustration at left].

"A refrigerator compressor can serve as the vacuum pump. To prevent oil vapor from back-streaming from the pump into the laser a filter should be in-serted between the inlet port of the com-pressor and the gas outlet of the laser. An adequate filter can be made by pack-ing a one-gallon glass jug with glass wool. Close the jug with a two-hole rub-ber stopper. Gas enters the filter by way of a tube that extends through one per-foration of the stopper to the bottom of the jug. Filtered gas flows through a short tube at the top of the jug that con-nects to the inlet of the pump. The jug is enclosed in a wood box to minimize the hazard of flying fragments if the jug accidentally implodes. The refrigerator compressor I originally used worked well but became excessively hot after several hours of continuous operation. At present I use a conventional vacuum pump.

"The dimensions of the closed-end manometer are not critical. The instru-ment is made of standard-wall eight-millimeter glass tubing. The scale gradu-ations are equal in millimeters to the quotient of the density of mercury (13.55) divided by the density of the vacuum oil. If the density of the oil is not known, it can be determined with suffi-cient accuracy by weighing a known volume. The oil should be degassed by keeping it in a vacuum for an hour or so before filling the manometer.

"The tube can be filled with oil most conveniently by exhausting it to a pres-sure of $10^{-4}$ torr and admitting enough oil to completely fill the closed arm when air is let into the open arm. Al-ternatively the end that is to be closed can be softened, pulled to a constriction and cut off at the narrowest zone, like the tip of a medicine dropper. Enough oil can then be sucked into the tube to fill the long arm to the tip of the con-striction. The tip can be sealed with epoxy. The accuracy of the measure-ments depends on the quality of the vac-uum created when the oil separates from the closed end of the instrument. Even a tiny bubble above the oil can introduce a significant error.

"The performance of the laser de-

*Manometer for measuring gas pressure*

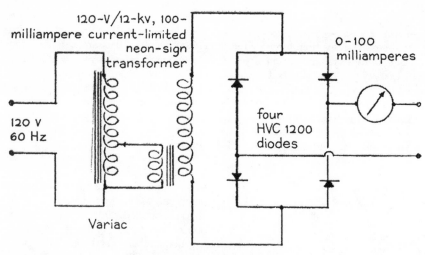

120-V/12-kv, 100-milliampere current-limited neon-sign transformer

120 V 60 Hz

Variac

0–100 milliamperes

four HVC 1200 diodes

*High-voltage power supply for the laser*

pends critically on the gas mixture, the gas pressure and the exciting electric current. An optimum mixture consists of eight parts of helium, two parts of nitrogen and one part of carbon dioxide. In the absence of helium the best performance was observed with a mixture of two parts of nitrogen and one part of carbon dioxide. The partial pressures are maintained at four torr of helium, one torr of nitrogen and .5 torr of carbon dioxide. These proportions and pressures assume that the diameter of the plasma tube is one inch. Experiments done with a plasma tube half an inch in diameter indicated substantially higher working pressures. With the narrower tube the optimum partial pressures ranged from 15 to 20 torr of helium, one to three torr of nitrogen and one to three torr of carbon dioxide.

"The electric-power supply consists of a variable transformer that feeds the input of a 12,000-volt, current-limited neon-sign transformer. The output of the high-voltage transformer is rectified by four silicon diodes connected in the bridge configuration [*see illustration above*]. The laser will operate satisfactorily on alternating current, but operation at maximum efficiency requires the use of direct current. If a conventional high-voltage transformer is substituted for the self-limiting neon-sign transformer, the output circuit must be equipped with a ballast resistor to prevent a runaway of current.

"The mirrors of the assembled unit must be adjusted to be parallel to each other and perpendicular to the axis of the plasma tube. To make the adjustment I first remove both mirrors by unscrewing their supporting mounting plates. The adjustment calls for three cardboard disks with 1/8-inch holes in

the center. Two of the disks are pressed lightly into the holes of the adjustable faceplates. A parallel beam of light is projected through the third disk, which is placed between the light source and the first aperture. The light beam is then directed through the holes of all three disks. The diameter of the beam should match the diameter of the holes.

"A helium-neon laser provides an ideal alignment beam. If such a laser is not available, an adequate beam can be formed by making a pinhole aperture in the slide carrier of a 35-millimeter projector and focusing rays from the pinhole into a parallel beam with a small telescope of the Galilean type. Place the light source on a rigid support at least 10 feet from the laser and adjust the position of the light source so that the beam just grazes the edges of the apertures. The light beam is then coaxial with respect to the plasma tube.

"Remove the two disks in the laser, leaving the third one in place. Install the concave mirror and adjust it to center the reflected beam on the remaining cardboard aperture. Repeat the procedure to similarly align the perforated mirror. When air is pumped from the tube, atmospheric pressure may distort some of the parts slightly and alter the alignment. This can be checked by leaving the collimating beam in place. When the system is in proper alignment, the collimating beam will be reflected by the output mirror and back through the third tube. If the reflected beam does not move when vacuum is applied, the output mirror is stable.

"The stability of the concave mirror can be checked by replacing the output mirror with a piece of flat glass. The system should now be in sufficiently good alignment for operation. After the

laser is oscillating the adjustments can be trimmed by trial and error for maximum power output.

"The operating procedure is fairly simple. Turn on the cooling water. Start the air pump and check the system for leaks. To make this test admit helium into the system to a pressure of 15 torr. Apply high voltage and adjust the current to approximately 100 milliamperes. During the first minute of operation the color of the discharge should turn from purplish to a pink-orange glow. The color change indicates that helium has replaced air inside the tube. Even a trace of the purplish hue indicates a leak in the system.

"A suspected leak can be confirmed by turning off the helium supply. Let the system pump down to below one torr. If no leak is present, the color of the discharge will look whitish gray. If any other color appears, turn off the current, tighten all sealing screws and check the gas-input system for leaks.

"When the system has been made gastight, exhaust the laser to the limit of the air pump. Then admit .5 torr of carbon dioxide, one torr of nitrogen and four torr of helium. Apply high voltage and with the variable transformer adjust the current to approximately 40 milliamperes. The laser should now begin to oscillate.

"The beam not only is invisible but also may be weak. It can be detected by inserting a small sheet of waxed paper or Thermofax paper in front of the output window. Caution: Do not place your hand or any part of your body in the path of the beam, even during the initial period of adjustment. The laser may be developing full output power, emitting a beam of sufficient energy to shatter glass many feet away. Even the reflected beam is hazardous. For this reason it is advisable to make a small container for disposing of unwanted beam energy. A metal box with a small opening is suitable. Coat the inside of the box with flat black paint and position the opening so that it intercepts the beam. The unwanted energy will be absorbed harmlessly as the radiation bounces around inside the box.

"To maximize the power output try small adjustments of the mirror-alignment screws, gas pressure, gas proportions and current. The output should be substantial, ranging from one watt to 10 watts. At optimum power the mode pattern of the beam will burn itself into a piece of wood. A microscope slide inserted into the beam when the laser is at optimum power will shatter."

# Infrared Diode Laser

# 6

*A solid-state laser made from
semiconducting materials*

March 1973

O f the numerous kinds of lasers that have been developed during the past decade, the smallest and least costly to assemble is the diode laser. The active element, a sandwich of semiconducting material the size of a pinhead, emits an intense beam of infrared radiation when it is energized by a direct current. The invisible beam holds promise as a carrier of various kinds of signals. It can also be employed for echo ranging and intrusion alarms and for demonstrating such aspects of wave behavior as refraction, diffraction, reflection and interference.

Diodes of gallium arsenide that are designed for generating pulses of coherent radiation have become available in recent months at prices that enable the experimenter to assemble a working laser for less than $30. Several diode lasers, which emit beams that range in peak power from four to 70 watts, have been built by Harry L. Latterman of Mesa, Ariz. He explains the apparatus and some experiments that can be done with it as follows:

"The diode lasers I have made consist of three subassemblies: the diode and its mounting; an electronic circuit that generates pulses of direct current, and a source of power. The smallest of my lasers, which can be made in one evening, emits 200 pulses per second. Each pulse persists for 50 nanoseconds (billionths of a second) and reaches a maximum intensity of four watts. When the laser is operated by a dry-cell battery, it is self-contained and portable. The apparatus can be mounted in a protective housing smaller than a pack of cigarettes.

"The active part of a typical diode is a crystal consisting of three or more distinct layers of semiconducting materials sandwiched between electrically conducting films of metal. The layered structure accounts for the electrical characteristics of the diode. All crystals consist of atoms bound together in a lattice by forces associated with electrons in the outermost orbits of the constituent atoms—the valence electrons that bind atoms together. In ordinary crystals the valence electrons are fully occupied. In effect they act as tie rods that constitute the structure of the lattice.

"A semiconducting material known as the $n$ type can be made by incorporating in the lattice a few atoms that have one valence electron more than the number that can fit into the lattice structure. The surplus electron can be detached from its parent atom rather easily. It can then migrate through the crystal as a carrier of current. Conversely, a semiconductor material known as the $p$ type can be made by incorporating a few atoms that have one valence electron less than the number that can be accommodated by the lattice structure. In this case a bond is missing from the crystal structure.

"An adjacent valence electron can become detached from its bonding site and drop into the missing bond. A hole then exists at the vacated site. In terms of its electrical behavior the hole has the properties of an elementary charge. Because it is deficient in negative charge, however, it acts as a positive charge. Crystals so made are known as $p$-type.

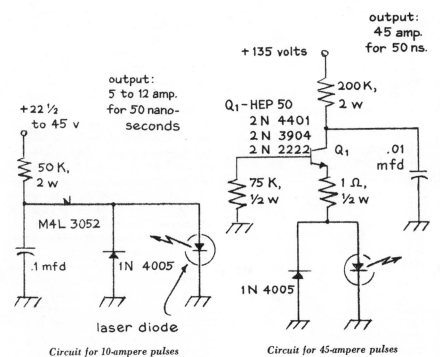

*Circuit for 10-ampere pulses*        *Circuit for 45-ampere pulses*

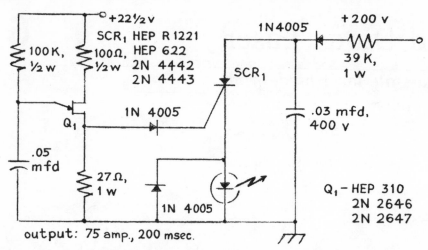

output: 75 amp., 200 msec.

*Circuitry employed by Harry L. Latterman for 75-ampere pulses*

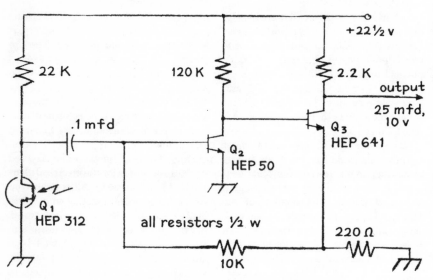

*Circuit for detecting infrared rays*

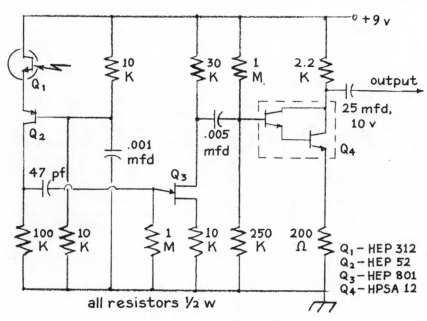

*High-sensitivity circuit for infrared detection*

"The crystal of a laser diode can be made with three layers of semiconducting material: a layer of $n$-type silicon, a layer of $p$-type gallium arsenide and a third layer of $p$-type gallium arsenide that also contains atoms of aluminum. The interface between $n$-type and $p$-type layers is known as a $p$-$n$ junction. Some easily detached electrons in the $n$ region migrate across the interface and drop into holes of the $p$-type material. Conversely, some holes in the $p$-type material cross the junction and wander randomly in the $n$-type material. The $p$-type material thus accumulates negative charge, and the $n$-type material similarly accumulates positive charge.

"The charges ultimately become strong enough to halt the migration of both holes and electrons. A potential barrier is then said to exist across the junction. The barrier can be modified by connecting opposite faces of the crystal to an external source of voltage. A negative potential applied to the $n$-type layer causes electrons to flow across the junction, migrate through the $p$-type material and return to the external source. The diode is said to be connected in the forward direction. Positive charge, as represented by the holes, migrates in the opposite direction. The action stops if the polarity of the external source is reversed.

"A positive charge applied to the $n$-type material simply increases the potential barrier. Hence the diode conducts current in only one direction. It is analogous to a hydraulic check valve and can be used to convert alternating current into unidirectional current, which is direct current.

"When the diode is in its quiescent state, before voltage from the external source is applied, the electrons have minimum energy. At room temperature the crystal is in a state of continuous but gentle vibration that causes electrons to wander aimlessly through the lattice. The interesting action begins when a forward potential of about 1.2 volts from an external source is applied across the $p$-$n$ junction. A current of electrons then flows through the diode. Collisions occur between the moving electrons and the electrons that are normally bound in the lattice structure.

"Some collisions are so violent that electrons emerge from the encounter with a discrete increment of additional energy. Energy so acquired is retained for a time. The excited electron is unstable, however, and it soon returns to a lower energy state spontaneously by

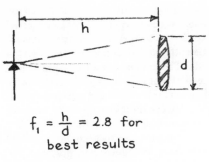

$$f_1 = \frac{h}{d} = 2.8 \text{ for}$$
best results

*Lens for infrared*

emitting a photon, or quantum of radiation, that carries away the acquired energy. Many electrons, acting independently, participate in the activity as long as the diode conducts current.

"The photons are emitted randomly in direction and in time. Hence the radiation is incoherent, analogous to the wave pattern created by tossing a handful of pebbles into a pool of still water. It is similar to the light emitted by a neon sign, and for the same reason. If a photon is pictured in the mind's eye as a short train of identical waves, coherent radiation can be pictured as a wave train consisting of two photons that unite in lockstep and proceed in the same direction.

"Coherent radiation can be generated in various ways. For example, it appears when a photon encounters and interacts with an electron that is on the verge of emitting surplus energy in an amount equal to the energy carried by the stimulating photon. The emitted photon

joins the stimulating photon and the two proceed through space together.

"In order to generate coherent radiation, lasers of all kinds provide two essential conditions. First, the laser must maintain an adequate supply of excited electrons. Second, the excited electrons must be trapped inside a resonant optical cavity that consists of a pair of facing mirrors.

"In the case of the diode laser an adequate population of excited electrons is generated by connecting a source of power to the diode in the forward direction to provide current in excess of a certain minimum threshold value. At currents below this value the diode emits incoherent radiation; such a diode is called a light-emitting diode, as distinct from a diode laser. The mirrors that form the optical cavity of the diode laser are merely the square-cut edges of the crystal. The edges function as mirrors because the index of refraction changes abruptly at the interface between the crystal and the air.

"Electrons that are excited in the region of the $p$-$n$ junction migrate into the transparent $p$-type material. Laser action begins when a photon is spontaneously emitted. The liberated photon bounces back and forth between the mirrors, interacting with the population of excited electrons and thereby stimulating the emission of additional photons.

"The interface between the $p$-type gallium arsenide and the $p$-type gallium aluminum arsenide, which is known as a

heterojunction, serves to confine the excited electrons and also to reduce the reabsorption of energy. In effect, the heterojunction improves the efficiency of the device. A portion of the accumulating radiant energy escapes through the mirrors in the form of a coherent beam, which is the output of the laser.

"The minimum current required to generate coherent radiation by the laser diodes now on the market ranges from about 10 to 80 amperes. The exact threshold is specified by the manufacturer, as is the peak current rating. The diode can be destroyed by current in excess of the peak value.

"The crystal in the smallest of my diodes is almost invisibly minute. A dozen such crystals could fit easily into the volume of a pinhead. Nonetheless, the diode is rated at a peak current of 10 amperes, which is equivalent to a current density in the diode on the order of 100,000 amperes per square centimeter! The problem of driving the laser at the required current without vaporizing the crystal is solved by using short pulses of current and by mounting the diode on a metal base that dissipates the liberated heat.

"The three pulsing circuits I shall describe are designed for generating peak currents of from five to 75 amperes that persist for intervals ranging from 25 to 250 nanoseconds. One or another of the three circuits will work with currently available diodes. The circuits draw current from the source less than 1 percent of the time. A duty cycle much greater

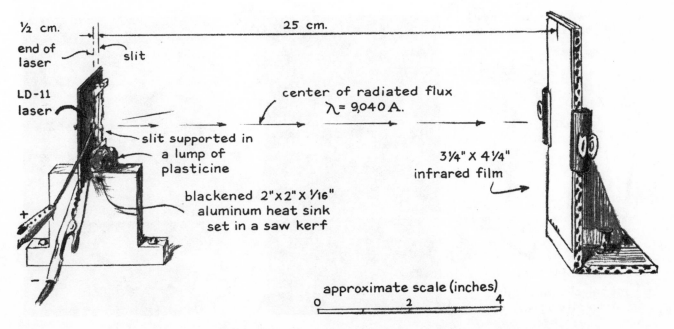

*Single-slit apparatus for demonstrating diffraction*

than 1 percent can damage the diode. The average drain on the power supply amounts to only a few milliamperes. The diodes are shipped in a protective housing that resembles a machine screw about half an inch long with a flat, cylindrical head a quarter of an inch in diameter.

"The infrared beam is emitted through a circular window of clear plastic in the exposed end of the housing. The diode can be fastened to a heat sink with the screw, which also serves as one of the electrical terminals. The other terminal, a short length of rectangular wire, is brought out of the housing. A two-inch square of sheet aluminum at least a sixteenth of an inch thick makes an adequate heat sink.

"A circuit that develops a peak current of up to 10 amperes for operating the smaller laser diodes consists of a resistor, a capacitor, a four-layer diode, a protection diode and the laser diode. Essentially it is an oscillator of the relaxation type. As the capacitor gradually accumulates charge through the resistor, voltage rises across the four-layer diode. At a critical potential the resistance of the four-layer diode falls abruptly. The capacitor discharges through the laser, after which the cycle repeats. The resulting pulse of current persists for about 50 nanoseconds, which is equivalent to a frequency of 20 megacycles. Hence the leads between the diodes and the capacitor should be made as short as possible.

"Magnetic fields develop around the

leads when the laser conducts current. At the conclusion of the pulse the fields collapse, inducing current in the wiring. As a result a reverse potential can appear across the diode. A reverse potential that exceeds a certain value can destroy the laser diode. The circuit includes a conventional diode that acts as a protective device, limiting reverse potentials to a safe value. Do not omit it. Peak current through the laser diode can be adjusted through the range from about five to 10 amperes by applying from 22½ to 45 volts to the circuit [*see illustration at left on page 35*].

"Laser diodes of intermediate power require peak pulses of current of up to 45 amperes. Currents of this magnitude can be developed by charging a capacitor to a potential of 135 volts. This potential is substantially higher than the breakdown voltage of the laser diode. Therefore an independent switch must be used for connecting the charged capacitor to the laser diode. The switching function can be accomplished by an appropriate transistor.

"A practical circuit is depicted by the accompanying diagram [*at right on page 35*]. A capacitor is charged through a 200,000-ohm resistor by the 135-volt source. As the capacitor accumulates charge, potential rises across the 75,000-ohm resistor in the base circuit of the transistor. When the capacitor reaches full charge, the potential of the base reaches a value that initiates conduction in the collector-emitter circuit of the transistor. The capacitor then discharges

through the transistor and the laser diode. A conventional diode, connected across the laser diode, limits the reverse potential and thus protects the laser diode. The circuit generates approximately 500 pulses per second. Each pulse persists for about 50 nanoseconds.

"Current pulses of up to 75 amperes are generated by charging a capacitor of .03 microfarad to a potential of 200 volts [*see top illustration on page 36*]. The charged capacitor is connected to the laser diode by a special switch in the form of a controlled silicon rectifier, a solid-state device that is capable of conducting relatively large currents. The switch is turned on and off by an oscillator, whose active element is a transistor of the unijunction type.

"The oscillator includes a 100,000-ohm resistor through which a .05-microfarad capacitor accumulates charge. Potential across the capacitor rises as charge accumulates and is applied to the base of the unijunction transistor. As full charge is approached the transistor conducts. The resulting discharge develops a voltage across the 27-ohm resistor in the transistor circuit. The voltage is applied through a diode to the gate terminal of the controlled silicon rectifier.

"Meanwhile the .03-microfarad capacitor in the laser-diode circuit has accumulated full charge from the 200-volt source. When the triggering voltage is applied to the gate of the controlled silicon rectifier, the rectifier conducts and the .03-microfarad capacitor discharges

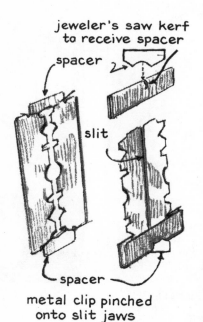

*Details of slit*

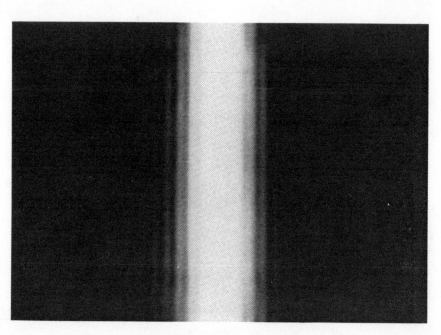

*Diffraction pattern from a single slit*

a 75-ampere pulse through the laser diode. Conventional diodes protect the laser diode from excessive reverse voltage. The pulse persists for about 200 nanoseconds.

"When I assemble any of the three circuits, I usually measure the pulse before connecting the costly laser diode in the circuit. The diodes are electrically equivalent to a resistor of about .1 ohm. I make up this resistor by connecting 10 one-ohm resistors in parallel. The voltage that appears across the resistor during a pulse is measured by an oscilloscope that can respond to a frequency of at least 200 megacycles. The current is calculated by Ohm's law: Current equals voltage divided by resistance.

"Several techniques are available for detecting the invisible beam of the lasers. The wavelength is quasimonochromatic and reaches peak intensity at about 9,000 angstroms. It turns out that the sensitivity of silicon phototransistors such as the type designated HEP 312 also peaks at this wavelength. I pick up the beam with this transistor and boost the resulting output signal by either of two simple amplifiers [see middle and bottom illustrations on page 36].

"The output of the amplifiers can be used for driving a power amplifier or a set of earphones or for triggering apparatus of other kinds. I use the two-stage amplifier in most experiments. The more sensitive three-stage amplifier is useful for detecting faint signals, particularly for picking up the beam at a substantial distance from the laser diode.

"The lasers emit the beam at a diverging angle of approximately 20 degrees. The angle can be reduced to about six minutes of arc by placing the diode at the focus of a simple lens with a focal ratio of $f/2.8$ or less. The aperture of the lens need not be large. An $f/2.8$ lens of half-inch aperture placed 1.4 inches from the diode is equivalent to and just as effective as a one-inch lens placed at a distance of 2.8 inches. Small, simple lenses of good quality are available from suppliers such as the Edmund Scientific Co. (Barrington, N.J. 08007) for about $1.

"The emission can also be photographed with conventional infrared film that is available from dealers who stock a full line of photographic supplies. To photograph the wave nature of the radiation, place the edge of a safety razor blade at a right angle to the beam and about one centimeter from the diode so that the edge intercepts about half of the beam. In a darkened room mount a sheet of infrared film in line with the blade and the diode at a distance of about 25 centimeters. The emulsion side of the film should face the laser. A laser diode of the M4L 3052 type will expose the film adequately in about two seconds. The accompanying photograph [preceding page] depicts the resulting diffraction pattern. A classical diffraction pattern can be similarly photographed by projecting the beam through a narrow slit, which can be improvised from a pair of razor blades.

"The cost of laser diodes varies with their size. Those of the smallest size have been advertised recently by dealers in surplus materials at a price of $5. The following companies can give additional information and prices concerning new diodes: Laser Diode Laboratories (205 Forest Street, Metuchen, N.J. 08840), Marketing Manager, Electro-Optical Devices, RCA Industrial Tube Division (New Holland Pike, Lancaster, Pa. 17604) and Texas Instruments Incorporated (P.O. Box 5012, Dallas, Tex. 75222).

"Finally, a word of warning: *The emission of all lasers, including that of diode lasers, is hazardous.* Infrared emission is particularly insidious because it is invisible and can be reflected by almost any smooth surface, including objects of tarnished metal and other materials that bear little resemblance to conventional mirrors. The beam can be transmitted safely for substantial distances only if it is directed through a metal pipe or a like enclosure. At shorter distances, up to a meter, the entire apparatus should be enclosed in an opaque box that will absorb the unwanted energy. Never operate the unshielded laser in the presence of other people."

# 7

# Nitrogen Laser

*An unusual gas laser that puts out
pulses in the ultraviolet*

June 1974

A recently developed laser that operates on a six-volt dry battery, emits 10 pulses of ultraviolet radiation per minute, each pulse about the size and shape of a broomstick. The pulses range in power from 50 to 100 kilowatts. They strike obstructions end on at the speed of light, with consequences that vary with the nature of the target materials. For example, the pulses bounce off clouds just as radar signals do.

With the echoes amateurs can measure distances to reflecting targets miles away; the accuracy is a matter of a few feet. With targets that absorb radiation the effects of impacts range from the emission of fluorescent light to the initiation of chemical reactions, including photochemical reactions. Indeed, as a source of radiation for making photographs the laser is about 10,000 times faster than the high-speed strobe lamps ordinarily used by amateurs.

Many parts of the laser can be assembled with materials that accumulate in the scrap box of anyone whose hobby

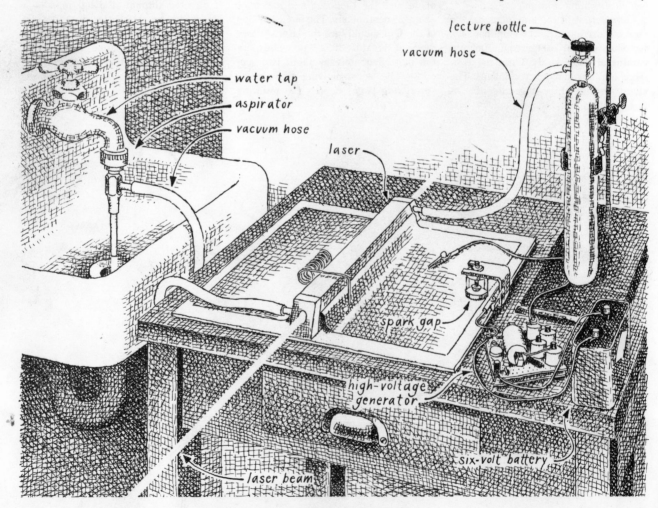

*James G. Small's nitrogen laser*

is electronics. A version of the apparatus that is particularly easy for amateurs to build has been developed and patented by James G. Small, a graduate student of physics at the Massachusetts Institute of Technology. Small explains the principles on which the apparatus is based and the details of its construction and operation.

"It has been known for some time that a high-current electric discharge in nitrogen gas that is flowing at a relatively low pressure can generate a pulse of coherent radiation, which is to say a laser gain, at the wavelength of 3,371 angstroms. The wavelength lies in the ultraviolet region of the electromagnetic spectrum where lenses and windows of most kinds of glass are transparent. The laser action begins when a molecule of nitrogen at room temperature absorbs energy by colliding with an electron that moves in the discharge. The encounter leaves the molecule in an unstable state. Usually it spontaneously falls to a state of lower energy by emitting a photon of radiation at 3,371 angstroms.

"The emitted photon may encounter another excited molecule of nitrogen and merely by its proximity stimulate the molecule to emit an identical photon. In this case the two particles of radiation join forces and proceed in the same direction, with their waves in lockstep. The resulting pulse of radiation contains twice the energy of each photon. This is laser action.

"The action will continue as long as the growing pulse encounters more excited molecules of nitrogen along its path than it does absorbing molecules. The process soon stops, however, because when a large number of molecules are suddenly excited, they will begin to randomly cascade to lower states of energy. Unfortunately in the case of nitrogen the molecules on the average linger at that lower level longer than at the upper one before moving on to still lower states. The number of molecules at the lower laser level builds up rapidly, eventually exceeding the number at the upper level and terminating the amplification. In fact, the gas quickly becomes strongly absorbing to 3,371-angstrom emission. The laser turns itself off even though there are still excited molecules left. Nitrogen lasers are therefore said to be self-terminating. The turnoff time is rather fast, usually less than 10 nanoseconds (billionths of a second), and it is responsible for the extremely short output pulse useful for radar and very-high-speed photography.

"The trick of inducing laser action in nitrogen lies in constructing a mecha-

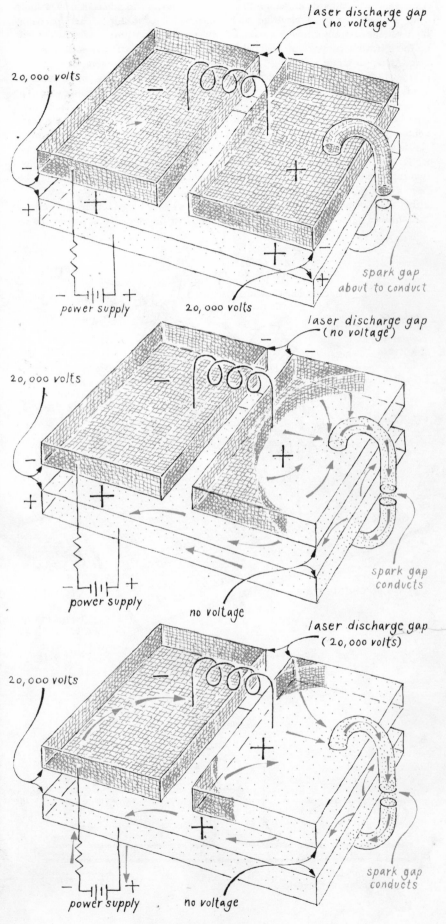

*The Blumlein switching phenomenon*

nism that will almost instantaneously send a huge current of electrons at a high voltage laterally through a column of the gas at a pressure of about 100 torr. An appropriate switching mechanism, which can handle tens of thousands of amperes within nanoseconds, turns out to be quite simple both in principle and in construction. It was invented by Alan Dower Blumlein, a British electronics engineer.

"Essentially the device consists of two adjacent metal plates separated from a third plate of equal total area by a thin sheet of plastic insulation [*see illustration at right*]. In effect the assembly behaves as an adjacent pair of interconnected capacitors. The space between the capacitors serves as the gap across which electric current can be discharged through nitrogen.

"The capacitors are interconnected electrically by a coil of copper wire. The capacitors can be charged by applying a potential difference between the interconnected plates and the mutual plate. Both capacitors charge to the same potential and the same polarity. No potential difference exists across the gap between them.

"The switching action devised by Blumlein develops when one of the capacitors abruptly discharges. The action is initiated by the breakdown of a spark gap that connects one adjacent plate to the mutual plate. Current moves through the spark gap when the charge accumulating on the capacitor assembly exceeds a predetermined voltage.

"As the charge rushes through the spark gap a steep difference of potential appears within the plate across a narrow boundary that separates the charged and discharged regions of the metal. The boundary has the form of a circular wave front that recedes from the spark gap at nearly the velocity of light. Shortly after the onset of conduction at the spark gap the voltage wave arrives at the center of the discharge gap between the plates. At this instant a potential difference appears across the center of the gap. Thereafter the advancing voltage spreads to the edges of the gap. In an apparatus 12 inches wide and 18 inches long a potential appears across the full length of the gap in less than .2 nanosecond, rising to its maximum value in about a nanosecond.

"If the discharge gap is enclosed by a container of nitrogen gas at low pressure and the capacitors are charged to 20,000 volts, the resulting discharge will raise an enormous number of the nitrogen molecules to the excited energy state. Laser action follows during the next five to 10 nanoseconds. The coil of copper wire through which the capacitor assembly is charged responds sluggishly to changes in current. For changes that occur within nanoseconds the coil acts as an open circuit.

"A practical nitrogen laser includes the switching apparatus, a power supply and a source of nitrogen gas, preferably of the grade used by welders [*see illustration on page 40*]. The compressor from an old refrigerator can be connected backward to operate as a vacuum pump for reducing the pressure of the nitrogen. An inexpensive water aspirator of the kind employed by chemists for vacuum filtering will work equally well.

"The capacitors of my laser were made from Type G-10 epoxy circuit board, which is clad with copper on both sides. This commercially available material serves widely in the electronics industry for interconnecting electronic apparatus. The unwanted metal is removed by etching to leave a network of conducting strips.

"My capacitors were formed on a single piece of circuit board that measures 30 × 45 × .04 centimeters (12 × 18 × .015 inches). Copper was etched from a two-centimeter (3/4-inch) margin

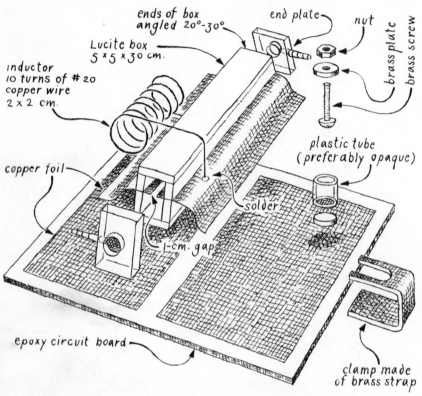

*Exploded view of the laser*

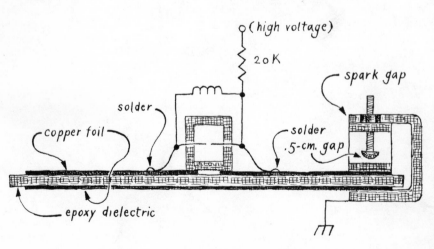

*The apparatus in elevation*

around the edge on both the top and bottom sides of the board. An additional strip five centimeters (two inches) wide was removed straight across the top side to form the two adjacent plates of the capacitor, each of which measures 18 × 26 centimeters.

"Copper surfaces that are to be preserved can be masked with waterproof tape or with the enamel spray paint that is called 'resist' by the photoengravers who developed this etching technique. The unwanted metal is preferably etched with ferric chloride, although dilute nitric acid can be used. In the U.S. all materials required for building the laser, including the etching chemicals, are available from North Country Scientific (R.F.D. 1, Plymouth, N.H. 03264).

"The capacitors can also be fabricated from a sandwich of metal foils or plates bonded to a thin insulating sheet such as acetate or Mylar and even certain paints made with plastic resins. The insulating material should be as thin as possible but should not conduct or break down when the maximum potential difference is applied across the plates. Epoxy circuit board can be operated reliably at a potential of 1,000 volts per .001 inch of thickness.

"The housing for the discharge consists of a gastight box five centimeters square and 30 centimeters long (outside dimensions). It can be made by cementing together strips of clear plastic, such as Lucite, .63 centimeter (1/4 inch) thick. An effective cement is clear silicone seal, a rubberlike material that is available at most hardware stores.

"The ends of the box are cut at an angle of from 20 to 30 degrees to prevent the reflection of radiation from the surfaces of glass windows back into the plasma, where it would be further amplified. The end plates are made of plastic that is thicker than the plastic on the sides. The added thickness makes it possible to drill holes in the edges for hose connections. These holes join other holes that pass through the center of the plates and serve as outlets for the radiation.

"The output ports are closed by windows made of glass microscope slides. The windows can be sealed to the plastic with paraffin wax of the kind used in home canning. The wax can be melted and applied safely with a heat lamp. The windows could be cemented in place with silicone seal, but wax has the advantage of reversibility (an advantage that will be appreciated when the glass becomes coated with grease from the vacuum pump or collects other foreign substances).

"The discharge electrodes of copper foil are cemented between two strips of plastic, which form each side of the box, with the edges of the foil opposed and separated by about a centimeter [*see bottom illustration on opposite page*]. The portion of the foil that extends outside the box has the form of a lazy S. The outer edge must be well soldered throughout its length to the foil of the epoxy board because the joint must conduct thousands of amperes.

"The spark gap must transmit maximum current in minimum time. It must be made of wide strap metal instead of wire to minimize its inductance. I made the clamps of the assembly with strap brass because of that metal's good conductivity, but aluminum strap could be substituted. An open air gap will work satisfactorily, but it is noisy and generates bright flashes of light that are rich in ultraviolet emission. I recommend that the gap be enclosed by an opaque tube of plastic or some other material that absorbs at least 90 percent of the emitted radiation.

"The laser must be charged to a potential of from 10,000 to 20,000 volts. A power supply that delivers .001 ampere of direct current will charge the capacitors to the ionizing potential of the spark gap several times per second. An alternating-current supply, such as a current-limited, neon-sign transformer rated at an output of 20 milliamperes, would cause the gap to conduct at twice the line frequency (120 discharges per second). Operation at this rate, however, would necessitate some means for cooling the gas if the maximum peak power is to be obtained. Warm nitrogen at low pressure does not 'lase' well. The gas should be allowed to cool to approximately room temperature after each pulse.

"The output is at a maximum when the discharge occurs in relatively pure nitrogen. The gas need not flow continuously. The discharge housing can be flushed with gas, pumped down to 100 torr and sealed off. The pulse-repetition rate of the sealed unit must be less than one pulse per second.

"Pulses at this rate can be generated with a battery-operated power supply [*see illustration above*]. This device consists of a low-voltage transistor oscillator, a step-up transformer and a voltage-multiplying rectifier. The transformer is made from an automobile ignition coil. Carefully cut the top from the housing of the coil. Remove the windings, along with the iron core. The innermost winding in most coils will consist of approximately 25,000 turns of fine wire. Carefully unwrap from the fine wire the outer, heavier-gauge primary winding and in its place wind two 20-turn coils of 20-gauge enameled wire, which should be closely spaced.

"Connect one of the coils in series with the collector of the 2N3055 transistor or an equivalent transistor such as the 2N3236, 2N5039, 2N6271 or HST9203.

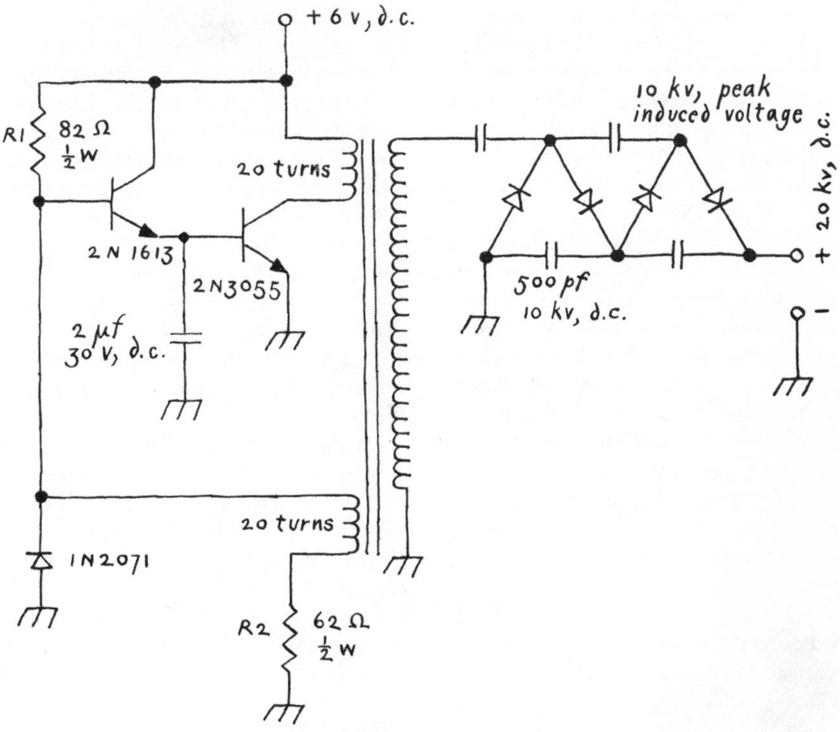


*Circuitry of the transistor power supply*

This transistor needs no heat sink, as it is operated conservatively. The 2N1613 is a low-powered transistor. It is connected in the Darlington configuration to speed the switching time of the larger transistor. Other transistors that will work nicely for this purpose include the 2N696, 2N2222 and 2N3642.

"The circuit is grossly overdesigned. Experimenters outside the U.S. will find that almost any silicon NPN transistor with a beta of 30 or more and a power dissipation of more than half a watt can be substituted for the 2N1613. For the larger transistor try any NPN of the silicon type with a collector current rating of five amperes or more that is housed in a TO-3 case. Surplus dealers in the U.S. have been selling large transistors of this kind at four for $1. When transistors are substituted, the values of resistors $R_1$ and $R_2$ may have to be altered experimentally within a factor of three to ensure reliable oscillation. The frequency of oscillation should be a few kilohertz.

"The power supply will operate from a six-volt battery at from 1.5 to three amperes and will deliver 20,000 volts at current sufficient to charge the laser to ionizing potential in from three to eight seconds. The electrical efficiency is quite low (about 1 percent), primarily because of the inefficient but simple transformer. Do not omit the two-microfarad capacitor that connects from the chassis to the junction between the emitter of the first transistor and the base of the second transistor. Without this capacitor the unit may tend to oscillate at a frequency too high for efficient transformer operation, depending on the loading of the high-voltage output.

"The reader may wonder why no mention has been made of laser mirrors. The optical gain of the rapid discharge is so large that emission becomes superradiant, which means that the unit will lase without an optical cavity. Radiation that is spontaneously emitted from molecules at one end of the laser can be amplified so strongly by the time it reaches the other end that the laser approaches saturation, meaning that it reaches the limit of its amplifying ability. Output is emitted from both ends of the column of excited gas, but a mirror at one end will more than double the power at the other end.

"The pulses can be detected easily by holding a piece of bleached cloth in front of either window. The cloth will fluoresce brightly. Almost any bleached cloth will function as a fluorescent screen. In general anything that glows in 'black light' works well: shirts washed in detergents containing fabric brighteners, 'psychedelic' posters, some kinds of paper such as white business cards, some types of clear mineral grease, the radiant dial of a watch or a clock and, of course, any of the dyes for dye lasers (fluorescein, relatively concentrated solutions of rhodamine 6G and so on).

"If the ultraviolet pulses are focused by a cylindrical lens to a line on the surface of the dye rather than to a point, the dye will often lase superradiantly in visible light along the direction of the line. No optical mirrors will be needed. Indeed, the nitrogen laser makes an ideal 'pump' for the dye laser. When it is employed as a pump, it opens up most of the visible spectrum to new experimental investigation [see "The Amateur Scientist," SCIENTIFIC AMERICAN, February, 1970].

"Rumor has it that some types of Day-Glo plastic will lase when pumped with a sufficiently intense ultraviolet pulse. I have seen motion pictures of this effect but have not done the experiment myself. The smooth surfaces of the plastic appear to function as the cavity of the optical laser.

"The ultraviolet laser can readily be scaled to higher powers. A discharge path one meter long can develop an output pulse of almost a million watts, although there is a trick to it. Because the laser turns itself off so quickly, radiation does not have time to travel the full length of the column before the gain automatically drops to zero.

"This problem can be solved with a traveling-wave discharge. Move the spark gap to a corner of the capacitor. The voltage wave will then arrive first at the end of the discharge channel nearest the spark gap and will race down the channel in step with the growing pulse of emission!

"As I have mentioned, all these lasers work best when the discharge channel is filled with flowing nitrogen at low pressure. Helium can be added with almost no effect other than raising the total pressure. With a sufficiently high percentage of helium the laser will work at atmospheric pressure, thereby eliminating the vacuum pump.

"During the development of this apparatus I have enjoyed the cooperation of Norman Kunit and other members of Ali Javan's laser group at M.I.T. I want to thank them for their spirited encouragement. I am particularly grateful to our chemist, Ray Mariella, Jr., for his suggestion of the aspirator vacuum pump.

"Finally, the experimenter must keep in mind that this is a high-powered apparatus and therefore a hazardous one. The ultraviolet emission from the laser and from the unshielded spark gap can harm the eyes. Avoid looking directly into the beam of this laser or any other one, just as you would avoid looking directly at the sun or at the arc of an electric-welding rig. Do not touch the capacitors until they have been completely discharged. Indeed, before touching any part of the apparatus make it a habit to short-circuit the spark gap with a yoke of wire supported at one end of an insulating rod about a foot long. (It is a good idea to cover the exposed high-voltage surfaces with sheets of Lucite.) Also keep in mind the fact that coherent energy, like sunlight, can be hazardous both in the direct beam and on a bounce as a specular reflection from a mirror or a smooth metal surface. Never project direct or reflected pulses into places where there may be people."

## Note on Extracting Nitrogen from Air

*October 1974*

Paul R. Burnett (2401 32nd Street S.E., Washington, D.C. 20020) suggests a simple procedure to extract nitrogen from air for use in the laser described in this department in June. Simply burn natural gas, propane or even alcohol and let the aspirator of the laser pull the combustion products successively through a bed of lime and a bed of anhydrous calcium sulfate respectively. Nitrogen of adequate purity will be drawn into the laser. A cold trap chilled with dry ice and acetone or any other desiccant can be substituted for the calcium sulfate.

# HOLOGRAMS

**II**

# II HOLOGRAMS

## INTRODUCTION

Holograms are three-dimensional photographs that offer a viewer the same advantages of perspective and parallax as the original scene. If you look into one of these photographs and then shift your head to one side, you will see the scene in the photograph from a new perspective. In effect you can look around the corner of an object in a photograph. This is impossible with a conventional, flat photograph where, if you shift your head, you still have the same perspective on the objects in the photograph.

Holography was once heralded as the photographic technique that would revolutionize media, including film and television. The revolution is yet to come, largely because of the technical difficulties involved in producing good holograms. Consequently, amateurs can still contribute to the new technology. The first steps will be to understand and duplicate the basic procedures for producing holograms. Most of these procedures are covered by Stong in the following articles.

Holography is based upon the interference of light. A beam of coherent light from a laser is split, with one portion, the reference beam, reflected by one or more mirrors until it falls on a photographic plate. The other portion is reflected by a mirror to the object that will be photographed. Some of the light scattered from the object illuminates the photographic plate and interferes with the reference beam. The interference pattern recorded on the plate encodes the information about the object photographed. When the plate is developed, it shows no recognizable pattern, but when coherent light is sent through the plate, replacing the reference beam, the interference pattern on the plate recreates the pattern of light produced by the object. As you peer into the plate and receive some of the light from the pattern, you see a virtual image of the object that had been photographed. When you shift your head to the side, you intercept different rays of light from the plate and gain a new perspective on the virtual image. The image provides the illusion of depth and parallax and thus simulates the original scene.

The requirements of a stable laser and a stable platform for the photography are essential. As an alternative to following Stong's suggestions for building a stable platform, you may want to construct a sandbox arrangement in which a sturdy box of sand is mounted on inflated innertubes. The large mass makes the structure stable, while the innertubes isolate the box from the floor's vibrations. The various optical pieces are mounted on cardboard tubes that are stuffed into the sand. Alignment of the pieces is easily accomplished by tilting the tubes in the sand. Instead of a box, you could substitute a bathtub, as did one finalist in the 1979 International Science and Engineering Fair. The products could then be called "bathtub holograms."

After mastering the basics of holography, you can proceed to an understanding of the more recent developments in the technology. The basic hologram described above is called a transmission-type hologram because when it is viewed, light passes through the photographic plate to the observer. One can also make a reflection hologram, in which light is reflected from the plate. With a special arrangement one can also produce a cylindrical hologram. Mounted much like a lamp shade, this type of hologram affords the observer any perspective of the object since the object is photographed in a full circle simultaneously.

Holography is on the verge of becoming a widely appreciated form of entertainment. After a few technological breakthroughs, it will be able to compete with television and other modern forms of entertainment. Capabilities to produce clear pictures, to record motion (more than is currently available in holographic movies), and to project full color must be developed. You may be able to contribute to these breakthroughs.

# 8

# Homemade Hologram

## Experimenting with homemade and ready-made holograms

*February 1967*

A few of the many amateurs who made the helium-neon laser described in this department in September, 1964, and December, 1965, have now taken up holography: the photographic technique of recording light waves in mid-flight and in effect reconstructing them later so that they continue onward unchanged. If the recorded waves are reflected by an object, the reconstructed waves enable an observer to see the object in three dimensions, as though he were viewing it through a window. The reconstructed scene appears in three dimensions, with full perspective and all the effects of parallax. For example, background details that may be obscured behind a feature in the foreground can be brought into view simply by moving the head to one side or the other. Similarly, the eyes must be refocused when attention is shifted from objects in the foreground to those in the background [see "Photography by Laser," by Emmet N. Leith and Juris Upatnieks; SCIENTIFIC AMERICAN, June, 1965].

It is now possible for amateurs to buy a hologram and to perform some interesting experiments with it, a subject to which we shall return. Sylvain M. Heumann of South San Francisco, Calif., is one of those who do their own holography. Discussing both the principles and the procedures he uses for making holograms at home, Heumann writes: "In principle holograms should be easy to make. In practice they are not.

"The object to be recorded is placed on a solid platform and flooded with light from a laser. Light reflected by the object falls on a photographic plate that faces the object. The plate is simultaneously flooded by a second set of rays, called the reference beam, that is reflected by a mirror. The reference waves travel on a path that bypasses the object. After adequate exposure the plate is developed.

"No lens is used to form an image, and no image appears on the completed plate. Instead the emulsion records an abstract pattern of fine lines and whorls that may be roughly likened to a thumbprint. If rays of colored light are now directed through the hologram along the path of the reference beam, a new set of rays emerges from the back of the hologram. The new waves are in every respect identical with those that were reflected by the object. A viewer who sees them for the first time is likely to think he is being tricked, because the object looks so real.

"If the exposure were made with ordinary light, the photographic plate would merely blacken. Every part of the plate receives light from every point on the object and from the mirror. Laser light, however, is coherent: the waves are identical in length, and they proceed in step. At some points on the photographic plate the crests of waves reflect-

*Reconstruction of entire hologram at top of opposite page*

*Hologram made with a laser built by an amateur*

*Perspectives obtained by reconstructing corresponding segments of the hologram*

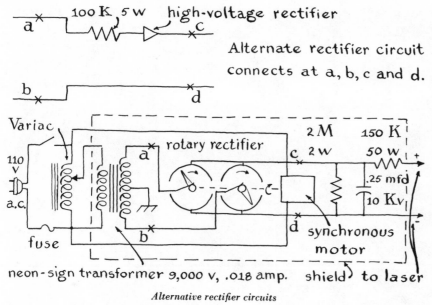

a ×———— 100 K 5 w —/\/\/\—▷—× c   high-voltage rectifier

Alternate rectifier circuit
connects at a, b, c and d.

b ×————————————× d

Variac

110
V
a.c.

fuse

rotary rectifier

a

b

2 M
2 w

c

150 K
50 w

.25 mfd
10 Kv

+

d   synchronous
motor

shield ) to laser

neon-sign transformer 9,000 v, .018 amp.

*Alternative rectifier circuits*

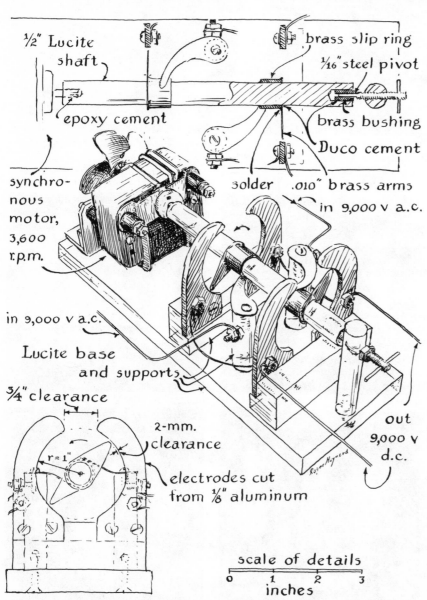

½" Lucite
shaft

epoxy cement

synchro-
nous
motor,
3,600
r.p.m.

in 9,000 v a.c.

Lucite base
and supports

¾" clearance

r = 1"

2-mm.
clearance

electrodes cut
from ⅛" aluminum

brass slip ring
1/16" steel pivot

brass bushing

Duco cement

solder  .010" brass arms
in 9,000 v a.c.

out
9,000 v
d.c.

scale of details
0   1   2   3
inches

*The synchronous-switch rectifier*

ed by the object coincide with the crests of waves in the reference beam. The two waves reinforce each other and expose the photographic emulsion at that point. At other places the crests of waves reflected by the object coincide with the valleys of waves in the reference beam. They cancel, so that the plate receives less exposure. Such interference effects vary at all points on the plate, depending on the shape and surface texture of the object.

"The pattern of fine lines in a hologram has the property of diffracting, or bending, light rays. The diffraction is greater with close spacing than with narrow spacing. Advantage is taken of this effect in the hologram to reconstruct the light waves that were reflected by the object. Rays that enter the hologram from the same direction as the reference beam are bent and scattered precisely enough to match those that were reflected by the object. In effect they duplicate the object rays. The light used for reconstructing the object rays should be coherent, but remarkable realism can be achieved with ordinary colored light emitted from a pinhole source.

"The structure of the hologram involves dimensions that are determined by the wavelength of light and the angle made between the object beam and the reference beam. Normally the plate must record many thousands of lines per inch, a resolving power that greatly exceeds that of ordinary photographic plates. Indeed, the ultrafine structure of the pattern explains why the hologram can record more information than an ordinary photograph. It also helps to explain why holograms are difficult to make at home. During exposure the photographic plate must remain motionless with respect to the relative positions of the mirror and the object. Any relative movement between the three in excess of a few millionths of an inch causes the lines to blur; hence the quality of the reconstructed waves will be seriously degraded.

"For best results the photographic plate must be capable of recording about 60,000 lines per inch. Emulsions capable of this high resolution are comparatively insensitive. Those I use are rated at an ASA speed of only .003, in contrast with ordinary black and white film, which is rated from 400 up. The problems of making holograms, then, consist in devising rigid structures for supporting the apparatus, insulating the apparatus against vibration and maximizing the available light to minimize the exposure interval.

"The first requirement for making holograms is a laser. Mine was built at home. The apparatus described previously in this department will work splendidly if it is modified to develop somewhat more power and to generate light waves of a single frequency. The output power can be increased substantially by operating the laser on direct current, a requirement that is simple to meet. A string of two or more silicon rectifiers, such as type CR210, can be connected in one lead of the neon-sign transformer as indicated by the accompanying illustration [*top, page 50*]. The resulting unidirectional current is smoothed by connecting a capacitor across the output of the rectifiers. The circuit must also include two resistors, one for limiting the current in the rectifiers and the second to compensate for the negative resistance of the laser tube. High-voltage rectifiers are expensive. The experimenter who has more time than money can substitute a synchronous rotary switch.

"The switch consists of an insulating shaft that carries two switch arms spaced 180 degrees apart. Each arm passes close to but does not touch an opposing pair of semicircular electrodes [*see bottom illustration on opposite page*]. In the case of 60-cycle operation the switch arm rotates synchronously at 3,600 revolutions per minute. It is driven by a Barber-Coleman synchronous motor, type KYAJ622–328. The motor is available from the Edmund Scientific Co., 101 East Gloucester Pike, Barrington, N.J. 08007. The base of the switch and the supports for the electrodes can be made of Lucite or any comparable insulating material.

"Alternating current is connected to the switch arms through brushes made of brass shim stock that ride on brass slip rings. The slip rings make a snug fit with the shaft. Other essential mechanical details are evident in the illustration. The inner diameter of the semicircular electrodes must be at least two inches, and opposing electrodes must be spaced at least three-quarters of an inch apart. All the other dimensions can differ from those shown.

"When the synchronous switch is in operation, the blades must stand midway between the opposing semicircular electrodes at the beginning of each cycle. They must complete half of a revolution at the end of each alternation of current. In other words, the switch must operate in phase with the alternating current.

"To set the switch arms in phase, connect one output lead of the neon-sign transformer to the brush of one switch

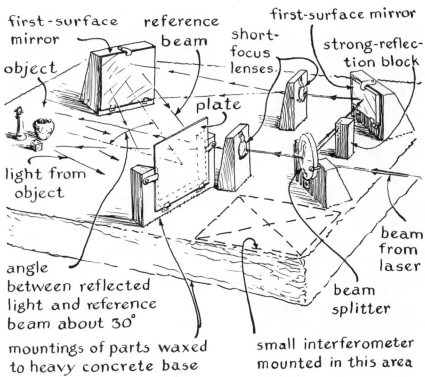

first-surface mirror reference beam
object
plate
light from object
short-focus lenses
first-surface mirror
strong-reflection block
beam from laser
beam splitter
angle between reflected light and reference beam about 30°
mountings of parts waxed to heavy concrete base
small interferometer mounted in this area

*Optical train for making holograms*

arm and connect the other output lead of the transformer to one of the semicircular electrodes. Apply power to the neon-sign transformer from a variable-voltage transformer, such as a Variac. Connect the motor to the power line. When the motor comes up to full speed and is running synchronously, gradually apply power to the neon-sign transformer and observe the gap between the switch arm and the electrode. When the voltage has been increased sufficiently, sparks will bridge the gap.

"Note the point on the semicircular gap where the sparks first appear. Perhaps they will begin approximately halfway around the electrode. If so, shut off the power, stop the motor and rotate the switch arm 90 degrees on its shaft. Reenergize the apparatus and again observe the gap. Doubtless the sparks will now fill the entire arc of the semicircular electrode. Should the sparks originate at greater or lesser angles around the semicircular electrode, adjust the angular position of the switch arm on its shaft by an appropriate amount.

"After the switch arm has been positioned so that the sparks fill the complete arc of the electrode, fix it to the shaft with a dab of quick-drying cement. Then rotate the remaining switch arm 180 degrees from this position and similarly cement it to the shaft. Adjacent semicircular electrodes are interconnected. When an alternating-current source is

connected to the rotating switch arms, unidirectional current can be drawn from leads connected to opposing semicircular electrodes.

"The switch functions as a full-wave rectifier and can replace the costly diodes. The switch requires no current-limiting resistor. A capacitor of about .25 microfarad should be connected across the output of the switch, however, and a resistor must be inserted in one lead between the capacitor and the laser to compensate for the negative resistance of the laser tube.

"Conduction between the switch arms and the semicircular electrodes is established through the spark. For this reason the switch unfortunately acts as a copious generator of electromagnetic noise at frequencies close to all television channels. In order to prevent the radiation of this noise, the switch, the neon-sign transformer, the capacitor and the resistor must be installed in a grounded metal cabinet.

"In some cases it may also be necessary to insert choke coils and bypass capacitors in the alternating-current power line and the direct-current output leads. The choke coils and bypass capacitors should be potted in grounded metal containers and installed in the cabinet. The cost of the complete synchronous rectifier should not exceed $20. Warning: The high voltage is lethal. Handle it accordingly.

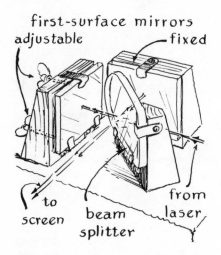

first-surface mirrors
adjustable — fixed

to screen   beam splitter   from laser

*Arrangement of interferometer*

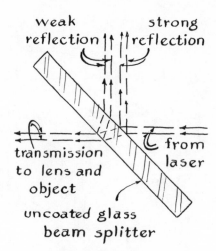

weak reflection    strong reflection

transmission to lens and object

from laser

uncoated glass beam splitter

*Details of the beam splitter*

"To make certain that the laser will generate coherent light of a single frequency (that it will operate in the so-called $TEM_{00}$ mode), the resonator should consist of one mirror of spherical figure and one flat mirror. The flat mirror can be bought from Henry Prescott, 116 Main Street, Northfield, Mass. 02118. The laser described previously in this department was equipped with a pair of mirrors of spherical figure. Either one of these can be replaced with the flat mirror.

"To make the modification, align the two spherical mirrors so that the laser functions normally. Remove one mirror and replace it with the flat mirror, which can be aligned by inserting a microscope slide between it and the adjacent Brewster window at an angle of about 45 degrees, shining a small light on the slide and manipulating the adjustment screws while looking through the spherical mirror and down the capillary tube. When the reflected light reaches maximum intensity, the flat mirror is in proper adjustment.

"The adjustment can also be made by the method described in this department in December, 1965. Occasionally a small additional adjustment is necessary. It is made by applying direct current to the tube and rocking the adjustment screws back and forth slightly until the beam appears. Direct the beam onto a white screen and observe the pattern. If it consists of an array of two or more spots, adjust the screws until the spots merge into a single disk of uniform intensity. Incidentally, the laser may not develop maximum intensity when adjusted for $TEM_{00}$ mode, but more intense multimode beams cannot be used for making holograms.

"The desired disk-shaped spot of light may contain a number of interference fringes and circles. Such spurious effects usually represent diffraction patterns that are caused by dust or by imperfections in the mirrors. The beam can be cleaned up by passing the light through a pinhole about .0005 inch in diameter.

The pinhole must be located at the focus of the two lenses that will be used to spread the beam into a pair of broad cones. A good pinhole can be made by pressing a sharp needle into a sheet of aluminum foil backed by a piece of plate glass. The pierced foil can be mounted on a ring of cardboard for clamping into position in the optical train. Finally, the power of the laser can be further increased 10 to 25 percent by placing a series of reasonably strong horseshoe magnets every inch or so along the laser tube. The magnetic fields reduce the tendency of the laser to generate infrared waves and therefore concentrate the output at the desired wavelength of 6,328 angstrom units.

"In addition to the laser, the experimenter will require the following equipment: a heavy table that is insulated against vibration; four first-surface mirrors; two lenses for spreading the laser beam; two beam splitters, and a supply of high-resolution photographic plates together with chemicals for their development. All these materials, except the table and the chemicals, can be bought from the Edmund Scientific Co.

"My table consists of a granite surface plate mounted on dense polyfoam. It weighs 100 pounds. The polyfoam rests on the cement floor of my basement. Another amateur who goes in for holograms uses a stack of concrete blocks of the type sold by dealers in gardening supplies. Each block is two feet square and two inches thick. Six blocks are cemented together with roofing tar and placed on a foot-thick stack of old newspapers. The assembled table weighs 500 pounds. The heavier the table the better. It cannot be insulated too well.

"To check the stability of the table you will require a small interferometer consisting of a beam splitter (Edmund catalogue No. 578) and two first-surface mirrors (Edmund catalogue No. 40,040). These components can be secured to one corner of the table by wax, blocks of wood or rigid fixtures such as machinist's vises [*see illustration at top left on this page*]. Direct the rays of the laser into the beam splitter and adjust the position of the components until the two beams superpose on a screen that can be permanently mounted on a distant wall.

"The superposed beams will make a small spot of light on the screen. Enlarge the spot by inserting a lens with a focal length of 10 to 50 millimeters in the beam at a point within a few inches of the apparatus. Interference fringes will appear in the enlarged spot. They must show no perceptible movement. If they

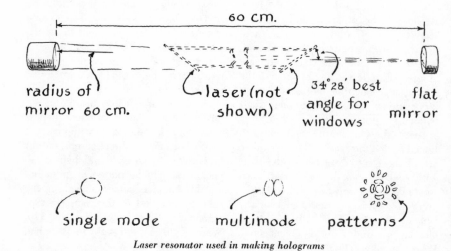

60 cm.

radius of mirror 60 cm.    laser (not shown)    34°28' best angle for windows    flat mirror

single mode    multimode    patterns

*Laser resonator used in making holograms*

do, add mass to your table and improve the insulation. During the hologram exposure the fringes must show no movement. Street traffic and other sources of vibration can present a problem. In some regions exposures can be made only during the early hours of the morning when traffic is at a minimum.

"Once the table has become stable you can assemble the optical train of the holograph apparatus [*see the illustration on page 51*]. You will require a piece of thick glass for the beam splitter (Edmund catalogue No. 2,183), a large front-surface mirror (Edmund catalogue No. 40,043), a small first-surface mirror (Edmund catalogue No. 40,040) and two simple lenses of good optical quality, any convenient aperture and a focal length of about 17 millimeters. The lenses need not be achromatic. The mounting supports can be improvised according to the tastes and resources of the experimenter. Again, stability is the essential requirement.

"The subject to be photographed should consist of small objects that will stand still. Chessmen are a good example. The available light from a homemade laser limits the size of the scene to about one square foot if the exposure is to be kept within a five-minute interval. The photographic plate should be placed vertically, facing the subject at a distance of about 10 inches. First, however, place a piece of white cardboard in the position the plate will occupy. The cardboard should match the size of the plate.

"Darken the room, direct the rays of the laser into the beam splitter and adjust the lens of the appropriate beam to floodlight the object. (The laser does not have to be on the stable table.) Block off this beam and adjust the lens and mirrors so that the second beam, as reflected by the small and large mirrors, floods the cardboard screen. If the diagram [*page 51*] has been followed carefully, the distance from the beam splitter to the object to the cardboard screen will be approximately equal to the distance from the beam splitter to the small mirror, large mirror and cardboard. In no case should an inequality exceed half of the length of the laser. If scattered light from the laser tube is perceptible on the screen, enclose the laser in an opaque housing.

"The two beams that now fall on the screen must be adjusted for relative intensity. The beam from the mirror should be two to three times brighter than the rays reflected to the screen by the object. The brightness is difficult to estimate, but it is not too critical. If the reference beam seems too bright, try shifting the position of the lens so that it picks up the rays that are reflected by the second surface of the beam splitter. If the beam still seems too bright, move the lens closer to the splitter or insert a neutral-density filter in the beam at the point where it is reflected from the beam splitter. If a filter is so used, place it exactly at right angles to the axis of the beam; otherwise light will be reflected back and forth internally between the glass surfaces and will introduce unwanted interference effects.

"The angle made at the photographic plate between rays from the object and those of the reference beam should not exceed 30 degrees. The spacing of the lines in the hologram varies inversely with the size of this angle and becomes so narrow at angles approaching 90 degrees that problems arise. Now replace the cardboard screen with the photographic plate. The emulsion side should face the object. (The emulsion side of a plate can be determined by the fact that it will stick to your lip.) The best emulsion for holograms is the Eastman Kodak Company's 649F, which comes in the form of four-inch by five-inch glass plates, packed 36 to a box. The plates are fairly expensive. They can be obtained from the Edmund company in smaller quantities. Other emulsions of lower resolving power can be made to work by using a narrow angle between the reference beam and the object beam. This arrangement generates somewhat broader fringes, which are better for such emulsions, but the adjustment is difficult and I do not recommend it to the beginner.

"Just before making the exposure, direct the laser beam into the interferometer and examine the fringes for movement. If they appear solid, switch the rays to the hologram beam splitter and make the exposure. A dim safelight can be used if it is kept at least 15 feet from the plate. The exposure time is a matter of trial and error. If the object is colored and not more than three inches in diameter, a laser output of five milliwatts should make an exposure of optimum density in about three minutes. If the plates are stored in a refrigerator, allow at least 30 minutes for them to reach room temperature before use. The Eastman Kodak Company recommends that 649F plates be developed for five minutes at 67 degrees Fahrenheit in Eastman H. R. P. developer. Thereafter the plates are fixed, washed and dried.

"The dried hologram can be inspected immediately for an image. Place the plate in the diverged beam of the laser at the angle of the reference beam. You should then see the object. Rotate the plate from side to side to find the angle that yields maximum brightness. Alternatively you can inspect the hologram by placing a filter of almost any color in the slide holder of a 35-millimeter projector and fitting a pinhole mask over the front surface of the projection lens. You can even use a flashlight of the penlight type if it is fitted with a self-focused bulb. When inspected by flashlight, the hologram will be fuzzy and the image will appear in the colors of the rainbow.

"If no image can be found, the probability is high that something moved during the exposure. Examine the plate under a microscope at a magnification of about 600 diameters. The pattern should consist of fine, crisscrossed lines. If these lines are not seen, some part of the apparatus certainly moved during the exposure. If the emulsion is much darker or lighter than a conventional photographic transparency, appropriately increase or decrease the exposure during

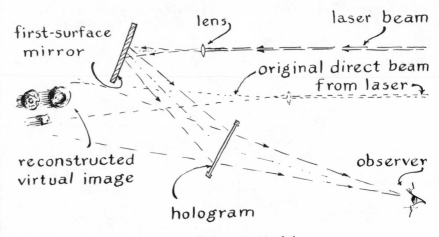

*Optical arrangement for reconstructing hologram rays*

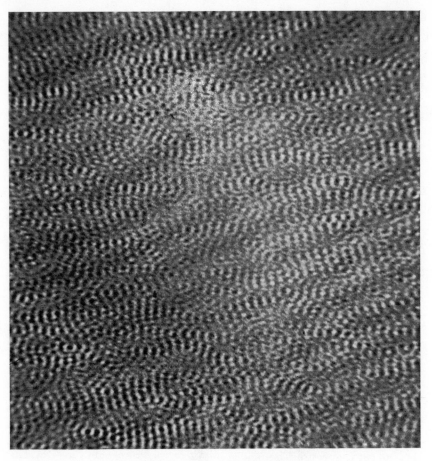

*Segment of a hologram enlarged 150 diameters*

the next try. If the object has poor contrast, increase or decrease the intensity of the reference beam.

"You do not have to make a hologram to have fun with these fascinating playthings. Relatively inexpensive holograms on film can now be bought, with a viewing filter, from the Edmund company. A number of engrossing experiments can be done with them. Try photographing the reconstructed light. You will discover that either the focus must be altered for recording sharp images of foreground and background objects or the lens must be stopped down to increase the depth of the field. Try making photographs of various areas of the hologram. Each portion, however small, will reproduce the entire scene—yet the information contained by each area differs from that of other areas. The principal difference involves the effects of parallax: the relative displacement of objects as seen from various points of view. In addition, each area makes a contribution to the resolution of the entire plate. Features of the scene appear in sharpest focus when the reconstructed rays from the full plate are intercepted by the eyes. It is this property of the hologram that accounts for the fact that blemished plates can yield good results. Clear photographs of objects can be made from holograms that are dust-pocked or scratched—imperfections that would ruin a conventional photographic negative.

"The hologram is a special form of the diffraction grating, which is a flat optical surface ruled with thousands of parallel, uniformly spaced lines and used for diffracting white light into its constituent colors. Diffraction gratings transmit part of the light as a straight beam but bend and disperse other portions into bundles of rainbow colors that lie on each side of the central beam. The bundles are known as diffraction 'orders.'

"The hologram also diffracts the light into such orders. If your eye is close to the hologram, you will find a certain angle at which an apparent mirror image of the scene appears. The depth of the field in the inverted image may appear greatly exaggerated, depending on the angle between the reference beam and the object beam at which the hologram was made.

"To gain a full appreciation of the astonishing amount of information that can be compressed into the two-dimensional pattern of the hologram, examine a plate under a microscope. At 40 diameters of magnification you will find curving lines making fine patchwork designs. At 200 diameters these fine details turn out to consist of still finer features. At 1,000 diameters the structures will be resolved into an orderly pattern of relatively straight, interwoven lines that resemble the seat of a caned chair. Good holograms contain more than 25,000 such lines per inch. It is in their number, shape and density that the optical information is encoded. These paragraphs have mentioned only a few of the many new optical experiments that have been made possible by the advent of the hologram."

# 9

# Stability of the Apparatus

*Insuring a good hologram by controlling vibration and exposure*

*July 1971*

Many amateurs have built a helium-neon laser. Relatively few have succeeded in using the laser to make a good hologram: the type of photograph in which the coherent light of the laser is used to create an interference pattern that, when it is illuminated with coherent light, yields a three-dimensional image of an object. Most of the failures can be traced to vibrations that produce minute changes in the distances between the various parts of the holographic optical system, particularly the distance between the object to be photographed and the film holder.

In the making of a hologram the coherent light of a laser is projected along two paths to the sheet of film. The rays traveling along one path constitute the reference beam; they fall on a flat mirror, from which they are reflected to the film. The rays traveling along the second path constitute the object beam; they fall on the object to be photographed, from which they too are reflected to the film. The rays of the reference beam are virtually equal in length. The rays of the object beam are reflected by the various surfaces of the object and are therefore of different lengths. As a result the light waves in the two beams are out of step and interfere with one another when they combine at the surface of the film. At some points on the film the crests of waves from the two beams coincide. The light is brightest at these points, and the film gets maximum exposure. At other points the crest of a wave in one beam coincides with the trough of a wave in the other beam. The light is dimmest at these points and the film gets minimum exposure. The resulting photographic pattern is essentially a record of the local differences in the length of the rays of the reference beam and the object beam. If the differences in path length so recorded arise solely from the shape of the object, the hologram yields a remarkably realistic image of the object. In practice, however, some portion of the difference invariably arises from the vibration of the apparatus. The quality of the hologram is degraded in proportion to the intensity of the vibration.

Professional workers who make holograms resort to elaborate schemes for isolating their apparatus from vibrations, such as mounting the parts on a multiton slab of granite, on a base that floats on air or on a platform equipped with a servo system that cancels the vibrations by means of feedback. People who do not have access to such apparatus can still make reasonably good holograms by means of a simple technique recently developed by LeRoy D. Dickson of the IBM General Systems Division in Rochester, Minn.

"In 1967," writes Dickson, "I found

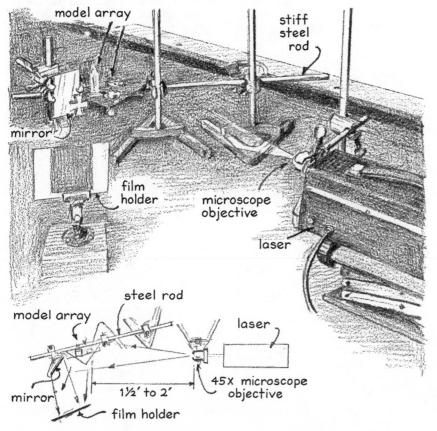

*LeRoy D. Dickson's setup for making a hologram*

that good three-dimensional holograms could be made by minimizing vibrations that alter the distance between the object and the mirror. Vibrations between the film and the laser, and between the laser and the object, have substantially less effect on the quality of the hologram, although they should be minimized. Essentially what I did was tie the two reflecting elements of the system together with a steel rod so that they moved as a unit. I attached the mirror to one end of the rod by an apparatus clamp and cemented the object to be photographed to a base of sheet metal that was in turn cemented to the rod a few inches beyond the mirror. The remaining portion of the rod was clamped to a pair of apparatus stands for support [*see illustration on page 55*]. This improvised optical bench was placed on a solid base that also supported the laser and film holder as separate units. A 45-power microscope objective lens was inserted in the beam of the one-milliwatt helium-neon laser to spread the narrow beam of the laser into diverging rays. The rays were directed toward the optical bench to simultaneously illuminate both the mirror and the object. The resulting holograms were sharp, bright and displayed good parallax.

"Two students at the Mayo High School in Rochester, Brandon Dallman and Tom Smyrk, experimented with the system and achieved excellent results on their first try. For objects they used mini-ature models of buildings and automobiles and mounted them as close to the mirror as possible. Indeed, one object, a Maltese cross, was mounted directly on the mirror. The one-milliwatt helium-neon laser, the microscope lens, the film holder and the rod assembly were placed on a pool table that rested on the concrete floor of a basement. I recommended the use of four-by-five-inch Kodak spectroscopic plates (Type 649-F).

"Light reflected to the film by the mirror should be approximately four times brighter than the light reflected to the film by the object. To achieve this ratio I use rough-surfaced objects of comparatively high reflectivity, such as unpainted models. The relative brightness of light that reaches the film can be judged by inserting a piece of white cardboard in the film holder and, with the room lights off, alternately blocking the rays from the mirror and from the object with a piece of black cardboard. Light reflected by the mirror can be reduced by moving the mirror either toward the edge of the beam or farther away from the laser. The best illumination ratio is 4:1, but anything from 3:1 to 10:1 will work. The angular position of the mirror should be adjusted to illuminate the film uniformly.

"To make an exposure, turn on the laser and let it warm up for 15 minutes. Turn off the room lights, block the laser beam with a piece of black cardboard, open the film holder and after a few sec-onds unblock the laser beam. The laser must of course be enclosed by a housing that blocks all light except the beam. Do not touch the apparatus with anything that might cause the parts to vibrate, including the cardboard used for blocking the beam. Try several exposures of 30 seconds, one minute and two minutes. Develop the films in Kodak D-19 developer for six minutes at 68 degrees Fahrenheit. Rinse, fix and dry according to Kodak instructions. Good holograms appear as transparencies that range from light gray to somewhat darker gray when they are examined in white light. The image can be reconstructed by looking through the hologram toward diverging rays from the laser. (It would be dangerous to look down the beam as it comes from the laser, but the diverging lens greatly reduces the intensity of the light.) Rotate the film around its vertical axis until the brightest image appears. The image can also be seen by looking toward a point source of ordinary light, but the quality will be degraded somewhat. A satisfactory source can be improvised with a 35-millimeter slide projector. Insert a piece of red gelatin in the space normally occupied by the slide and insert a pinhole mask over the front of the projection lens. The mask can be a piece of cardboard perforated by a hole about three millimeters in diameter. The perforation should be located at the center of the projection lens."

# 10

# Holograms with Sound and Radio Waves

## Sound and radio waves recorded on film by a precooling process

November 1972

Many an experimenter has hankered for a way to record sound waves and radio waves as conveniently as he can record light waves with a photograph. It now appears that all that is needed is a new method of processing a material that has been at hand for many years: Polaroid film. The method was discovered in 1968 by Keigo Iizuka, who was then a lecturer at Harvard University and is now associate professor of electrical engineering at the University of Toronto. His procedure opens up several new fields of experimentation, including microwave holography. Iizuka writes:

"Fields of ultrasonic waves and microwaves have ordinarily been investigated by two classical procedures. The wave pattern can be scanned mechanically by a probe that picks up a signal that varies in amplitude with the energy in each increment of the field. The resulting data are plotted to display the distribution of the energy. Alternatively, the field can be simulated in two dimensions by an appropriately shaped film of moving water. Streamers of dye in the water record the force potentials in the field [see "The Amateur Scientist," SCIENTIFIC AMERICAN, July, 1967].

"Both techniques are tedious and time-consuming. Moreover, the accuracy of the scanning technique can be degraded to the extent that the probe disturbs the field. The results of the fluid-mapping technique can be no better than the similarity of the fluid model to the field it represents. In contrast, Polaroid film can be employed to map the fields directly in less than five minutes. The method is both convenient and inexpensive. The intensity of a field can be measured simply by holding the film in the path of the waves.

"The procedure is based on the fact that the rate at which an emulsion develops after it has been exposed to light varies with its temperature or with the degree of mechanical agitation of the reagent. A microwave generates heat and an acoustic wave agitates the reagent. To record a wave pattern the emulsion is exposed uniformly to light of a certain color, precooled to an optimum temperature, coated with developing reagent and immersed in the pattern of waves to be recorded. The emulsion darkens at rates proportional to the local heating or agitation to form an image of the wave pattern. Development is interrupted when the darkest parts of the image reach maximum density.

"Polacolor Type 58 film, which is designed for making four-by-five-inch color pictures, is particularly appropriate for recording fields of sound and radio waves because the interval required for developing the resulting image varies with the color of the light to which the film is pre-exposed. The relation between development time and color arises from the structure of the negative. The negative consists of three major layers of dyed emulsion on a plastic base [see upper illustration on page 59]. After exposure to light the film pack is drawn between a pair of rollers that break a pod containing developing reagent in the form of a viscous jelly. This action also distributes the reagent as a thin film between the negative and positive emulsions.

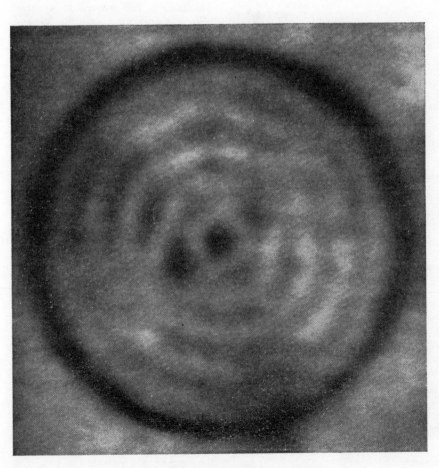

*Pattern of vibration of sound from a horn*

*Fringe pattern of 24.26-gigahertz microwave*

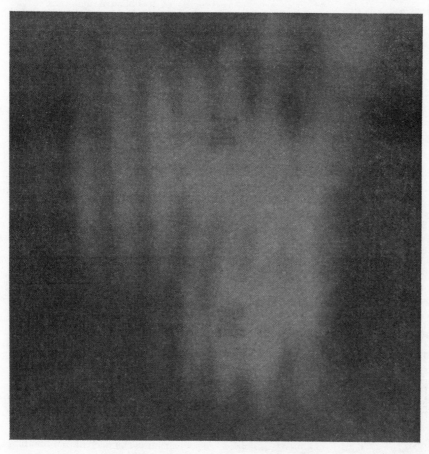

*Microwave hologram (M's are not holographic)*

"The layer of the negative emulsion that contains yellow dye is in intimate contact with the positive emulsion. The magenta and cyan layers lie progressively deeper in the sandwich. The time required for the reagent to reach each layer of dye and migrate to the surface for transfer to the positive emulsion varies with the depth of the layer.

"A negative emulsion that has been pre-exposed to yellow light develops in less time than one exposed to magenta or to cyan. For this reason I pre-expose with dark blue when photographing weak fields of microwaves (fields on the order of 60 milliwatts per square inch at a frequency of from one to 10 gigahertz). Conversely, pre-exposure to yellow light is used for relatively strong fields.

"Pre-exposures can be made with any camera that fits the Model 545 Polaroid four-by-five-inch Land film holder. The camera is focused on a screen of white paper that is uniformly illuminated from the sides by a pair of carbon arc lamps. I check the intensity by supporting a Kodak Neutral Test Card at the position of the screen. With an exposure meter I measure the light reflected by the card. The arc lamps are adjusted for a reflected intensity of approximately 50 footcandles. The camera is equipped with a holder for supporting Wratten light filters.

"To make a pre-exposure I set the lens opening at $f/9.5$ and adjust the shutter for an exposure of 1/2 second. I use a No. 13 Wratten filter for yellow and a No. 35 for magenta. Cyan is made by double exposure. In my experiments I first expose through a No. 47 Wratten filter (blue) for 1/10 second at a lens opening of $f/9.5$ and then through a No. 61 (green) for 1/5 second at the same lens opening.

"After making the pre-exposure I cool the film either with dry ice or with an instant-freeze aerosol can of Freon. During subsequent handling in air the film warms to a temperature of about 25 degrees Fahrenheit. That is the optimum temperature at which to make exposures of radio or sound waves.

"A small box in which the film can be refrigerated with dry ice can be made from a sheet of foam plastic. Make the box somewhat larger in area than the film. Cut a rectangle of plastic to serve as a cover. Put crushed dry ice in the box. Place the film on top of the ice. Close the box with the cover for a few minutes. Care must be taken to keep the portion of the film packet that contains the reagent pod outside the box. The reagent solidifies at 32 degrees F. Do not freeze it!

"Alternatively, the film pack can be cooled by spraying it with Freon. Aerosol cans of the refrigerant, such as those manufactured by the Cryokwik Company for freezing biological specimens, are available for about $1 per container from dealers in biological and pharmaceutical supplies. Avoid spraying the reagent pod.

"I have experimented with several types of black-and-white Polaroid film. Types 52, 57, 55-P/N and 107 work, but they are less sensitive than color film. In cases where interest is confined to a small area of the field the Polaroid eight-exposure color pack can be used to advantage. The image area of this film measures 3¼ by 4¼ inches. The pack is less convenient to use than the individual packs of four-by-five-inch film because it is difficult to cool the film without freezing the reagent pods that are encased in the pack. Microwave fields of larger area can be mapped with Polaroid Radiographic Packet Type TLX. This emulsion was developed for X-ray work, but it responds well to heat or agitation induced by microwaves or ultrasonic waves. The image area measures 9⅜ by 10½ inches.

"The pre-exposed and cooled film is placed in the Land film holder and pulled through the steel rollers that break the pod and spread reagent between the negative and positive layers. The pack is promptly removed from the film holder and inserted in the field to be photographed. In the case of microwaves the component of the electric field in the plane of the emulsion generates heat in the silver halide by means of induced current. The current raises the temperature of the emulsion in proportion to the square of the field intensity. A thermal field thus appears in the film. It is a replica of the intensity distribution of the electromagnetic field.

"The localized heating increases the localized rate at which developing reagent diffuses to grain sites of the silver halide and in addition accelerates the chemical reactions of the development. If development continued to completion, the film would turn uniformly dark. The chemical action is interrupted at an intermediate stage by stripping the positive emulsion from the negative emulsion. Development stops at once. The positive emulsion is then insensitive to light. The proper interval of exposure to microwaves or sound waves must be determined experimentally because it depends on the strength of the field.

"Most of my experiments have been made with microwaves, the field of my primary interest. In a typical experiment two beams of microwaves that vibrate at a frequency of about eight billion ($8 \times 10^9$) cycles per second are projected from a pair of horns. The beams cross at right angles and interfere at the zone of intersection.

"The film pack is placed in this zone at an angle of 45 degrees with respect to each of the beams [see top illustration on following page]. The pattern of standing waves that results from the interference is a grid made up of alternate strips of warm and cool emulsion. The developed image consists of alternately dark and light fringes comparable to the fringes that appear when two beams of light similarly interfere.

"I have made a microwave hologram of a small coin and a triangle of metal inside a leather purse. Three steps are involved in the technique. The purse is illuminated by microwaves that vibrate in the vertical plane. The waves are projected by a horn, as in the interference experiment. The energy is generated by a klystron oscillator at a frequency of 34.26 gigahertz (equivalent to a wavelength of 8.756 millimeters). The oscillator develops an output power of 10 watts.

"The beam of microwaves is directed at a right angle to the plane of the purse. The pack of Polaroid film, prepared by pre-exposure to cyan light, is placed directly behind the purse at an angle of 45 degrees with respect to the plane of the purse. The time of exposure to the

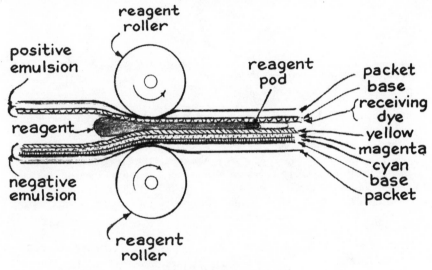

*Elements of Polaroid color-film pack*

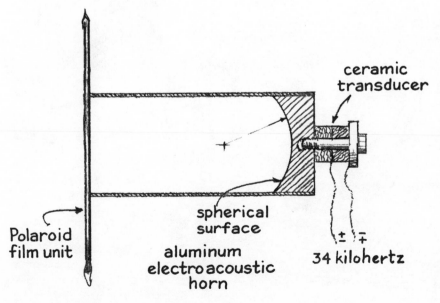

*Keigo Iizuka's apparatus for demonstrating interference of sound waves*

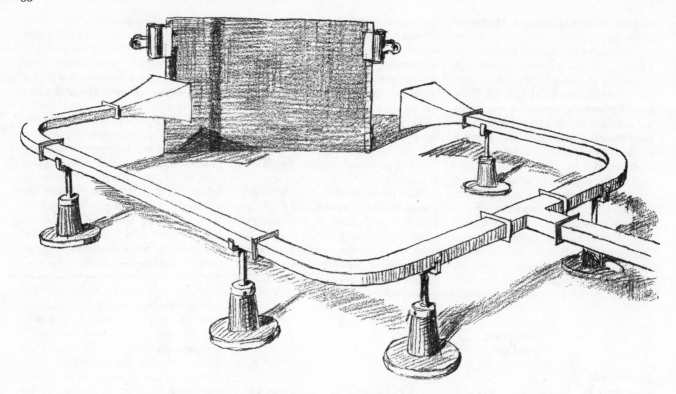

*Apparatus for demonstrating microwave interference*

microwaves ranges from 45 to 60 seconds.

"The developed image displays a series of interference fringes. They constitute a hologram of the purse and its contents. The hologram cannot be used directly for reconstructing the image of the object with microwaves because the positive print is transparent to electromagnetic radiation.

"To reconstruct the image a copy of the hologram is made in metal. I use aluminum foil reinforced by a cardboard backing. Slits are cut from the foil that correspond to the darkest portions of the fringes of the image, as judged by eye. A more accurate copy can be made in metal by substituting copper foil for the aluminum and utilizing the photoetching technique employed for making halftone engravings. Replicas made of aluminum foil by hand are adequate, however, for illustrating the procedure.

"The holographic image is reconstructed by illuminating the metallic hologram with the same beam of microwaves that was used to make the original hologram. The image can be made visible by either of two techniques. It will appear on a liquid-crystal film that is placed behind the metallic hologram. The liquid-crystal film consists of four layers: a Mylar sheet, a radio-frequency-absorbing layer, an active layer of cholesteric material and a top surface of polythene sheet. A water-soluble carbon

paint applied to the back of the liquid-crystal assembly was found to be suitable for absorbing radio-frequency energy.

"The reconstructed image is reasonably good but, as one might expect, the resolution is inferior to that of holograms

made with the far shorter wavelengths of coherent light. The image can also be made by substituting sensitized Polaroid film for the liquid crystal, although the film is less sensitive. Even so, the faint image is sufficiently recognizable to demonstrate that microwave holograms

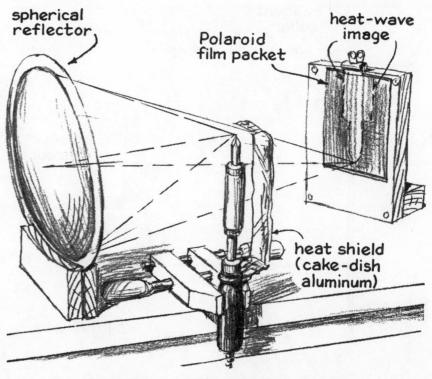

*Equipment for photographing an object by emitted heat*

can indeed be made by this technique.

"It is well known that a field of sound waves accelerates the development of photographic emulsions. Localized vibration in the emulsion increases the rate at which the developing reagent migrates to the sites of silver halide grains, thus speeding the chemical development in regions of high sound intensity. The effect can be demonstrated with any device capable of generating a loud monotone for a few minutes.

"Interesting patterns can be made in a small area with an ultrasonic generator. For example, I generate sound waves about a third of an inch long with a piezoelectric transducer of the ceramic type that is coupled to an electroacoustic horn [see lower illustration on page 59]. The transducer operates at a frequency of about 34,000 vibrations per second and develops a sound volume of 160 decibels at the mouth of the horn. The wave pattern is recorded by placing the sensitized film over the mouth of the horn. Clear images of the acoustic field resulted from exposures of 75 seconds for Type 58 Polaroid film and 10 seconds for Type 51.

"The minimum intensity required to register an image is about 80 decibels. Similar experiments have been made successfully at much lower frequencies. For example, I have mapped the field-intensity pattern inside an acoustic resonator that was generated by driving an ordinary loudspeaker with a signal of six watts at a frequency of 315 cycles per second.

"It is apparent that a pattern of wave interference could be recorded by superposing a reference field of sound waves on waves from a primary source. This method would be useful as a new means of visualizing acoustic fields as well as a possible method of making acoustic holograms. The relative insensitivity of the film to sound would seriously limit the usefulness of the technique, however, unless the experimenter had access to an acoustic source of high power.

"Numerous experiments can be undertaken to demonstrate the sensitivity of the pre-exposed films to heat. For example, the distribution of heat within the flame of a candle can be mapped by holding the sensitized pack vertically in the flame for a few seconds. The temperature distribution within the flame generates a corresponding distribution within the emulsion. To make a photograph of a candle flame I pre-exposed the film to magenta light. The bottom portion of the flame did not register in the image because it was in the margin of the film.

"A hot object, such as a soldering iron, can be photographed by focusing the heat rays on the film with a parabolic reflector [see bottom illustration on preceding page]. The contrast in the resulting image can be improved by inserting a heat shield between the hot source and the film. The shield prevents unfocused rays from reaching the sensitized emulsion.

"The presensitized film will respond even to mechanical vibrations. This effect can be demonstrated by attaching the film to a Chladni plate with fast-curing cement and striking the plate with a hammer [see illustration below]. The experiment works best with a plate of aluminum at least 1/4 inch thick. Regions in which the plate vibrates at maximum amplitude appear as dark bands in the resulting image."

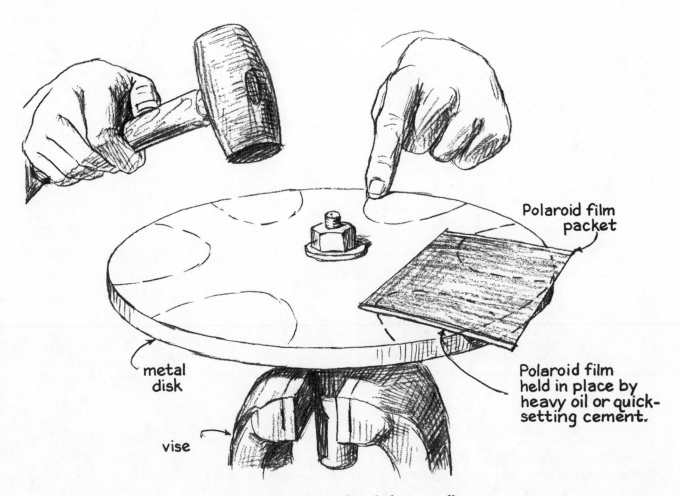

Polaroid film packet

metal disk

vise

Polaroid film held in place by heavy oil or quick-setting cement.

*A means of recording mechanical vibrations on film*

# INTERFEROMETERS III

# III   INTERFEROMETERS

## INTRODUCTION

Interferometers are instruments that use the interference of light to reveal optical parameters of microscopic scale. The instruments may be grouped in two classes. One type divides parts of a single wavefront into separate beams and allows them to interfere with each other. The other type divides a single beam into separate beams by means of reflection and refraction.

An example of a simple interferometer of the first variety is a card with two pinholes that is illuminated with a distant light source. The large distance between the card and the light source ensures that at any given instant there is a single wavefront striking the card. The two holes separate out portions of the wavefront. Because light passing through small apertures is diffracted, the emerging beams spread out. When they fall on a distant screen, the two beams overlap and interfere to give a characteristic pattern of bright and dark bands that correspond to constructive and destructive interferences.

One of the following articles illustrates another example of interference by the division of a wavefront. When a dull surface is illuminated with the coherent radiation of a laser beam, interference occurs at an observer's eye when a portion of the beam reflected at one place on the surface interferes with the light reflected at another place. Although the resulting interference pattern is far more complicated than that afforded by the simple interferometer with two small apertures, the basic phenomenon of the wave interference of light is the same.

The most important example of the second class of interferometer is the one constructed by A. A. Michelson near the turn of the century. In it a beam of light from a light source was made to fall on a thin plate of glass at such an angle that part of the light was reflected in one direction and another part was transmitted in another direction. Mirrors reflected the two light beams back to the glass where they then passed to the observer. When the instrument was properly adjusted, the two beams interfered with each other such that the observer saw a set of circular bands of constructive and destructive interference.

For the Michelson interferometer to be effective, the paths taken by the separated beams must be comparable in optical length; that is, the two beams must require approximately equal times of flight along the two paths. Consequently, if a transparent material such as a vial of gas or a clear layer of glass is inserted into one of those paths, the interference pattern will reveal the optical uniformity and the index of refraction of the intervening material.

Another method of splitting one beam of light into two beams is to have it fall on a thin film. Examples of the resulting interference pattern are numerous and include the colored bands seen in thin soap films. In such interferometers, a portion of the incident light reflects from the front surface of the film and then travels to the observer. Another portion of the incident light passes through the film, reflects from the back surface, passes through the film again, and then finally emerges to travel to the observer. The two beams of light interfere with each other at the observer to give the appearance of bright and dark bands across the film. Because the pattern depends on the thickness of the film, the effect can be employed to determine that thickness.

Interferometers can be used to measure the wavelength of light, to determine the optical uniformity of a transparent material, to measure the index of refraction of a transparent material, and to measure the linewidths and fine structure of the optical emissions from atoms and molecules. In what follows, we will address the problem of measuring the wavelength of light.

Determining in the laboratory the wavelength of common waves other than light, such as water waves, is relatively simple. One merely holds a measuring stick near the waves and notes the distance between the adjacent crests. How does one measure the wavelength of light? There is no convenient measuring stick for something so small. Furthermore, unlike other common waves, light involves no evident motion of matter. How does one measure the distance between the adjacent "crests" of the electric and magnetic fields of which the light is said to be composed? An interferometer, because of its extreme sensitivity to microscopic changes in lengths, is able to measure these wavelengths with great accuracy.

The wavelengths of the important emissions of common elements are now well known. Currently, the interferometer is employed to standardize the length of a meter in terms of the wavelengths of a particular emission from the isotope krypton-86. The procedure of calibration is complex, but in essence it involves counting the number of shifts from light to dark bands seen in a modified Michelson interferometer when one of the mirrors is moved a meter.

There are countless other examples of how interferometers have been useful in physical measurements, two of which are included in this collection. One involves measuring the speed of a moving mirror, and the other focuses on determining the dirt content of water. If you build your own interferometer, I am certain you can discover many more uses.

# 11 Michelson Interferometer

## A homemade instrument that can measure a light wave

*November 1956*

When two rays of light from a single source fall out of step (say after they have taken different paths and met at a common point), their waves reinforce or counteract each other, just as out-of-phase waves in water do. This effect accounts for the blueness of the bluebird, the fire of opals, the iridescence of butterfly wings and the shifting colors of soap bubbles. It also accounts indirectly for the accuracy of watches, the control of guided missiles, the quality of high-test gasoline and myriads of other achievements of technology which would not be possible without precise standards of measurement. All measurements, in the final analysis, depend upon one standard—length. Nowadays the length of the meter is calibrated in terms of wavelengths of light. The most commonly accepted standard for determining the length of the meter is the wavelength of the red light emitted by glowing cadmium in the vapor state. By interferometer methods this wavelength has been measured to a high degree of precision and comes out to be 6438.4696 ± .0009 Angstrom units (an Angstrom unit is one 10-billionth of a meter). The meter is 1,553,164.60 ± .22 times this wavelength.

The interferometer, the most elegant of yardsticks, is a singularly finicky and frustrating gadget. A scientist once remarked: "Without a doubt the interferometer, particularly the version of it developed by Michelson, is one of the most wonderful instruments known to science—when it is operated by A. A. Michelson!" In Michelson's hands the instrument certainly established an impressive row of scientific bench marks. It was he who measured the wavelength of the red cadmium line given above.

Making measurements with instruments capable of yielding precision of this order is not easy. One can fiddle with the controls of the interferometer for hours without seeing the fringes, or bands of interfering light, that serve as the graduations of length. No amateur would dream of making the instrument primarily for the purpose of using it regularly as a tool of measurement. But in constructing an interferometer and mastering the art of using it, one can learn a great deal about optics.

You can begin by repeating an experiment first performed by Isaac Newton, which demonstrates the basic principle. Simply press a spectacle lens against a glass plate and look directly into the light reflected by the combination from a wide source of light. If you use a magnifying glass, you will see several rainbow-colored rings, surrounding a tiny black spot about 1/64 of an inch in diameter at the point where the lens touches the plate.

The same effect can be observed with two sheets of ordinary window glass. An irregular pattern of interference fringes will surround each point at which the surfaces of the glasses touch. The pattern will be more distinct if the light source has a single color, *e.g.*, the yellow flame produced by holding a piece of soda glass (say a clear glass stirring rod) in a gas burner. If the glass sheets are squeezed even slightly during the experiment, the pattern of fringes will shift, indicating the minute change in distance between the inner faces of the sheets.

Thomas Young, an English physician, and his French colleague Augustin Fresnel demonstrated in the latter part of the 18th century why interference fringes appear. In so doing they established the wave nature of light. They explained that if two rays of light emitted from the same source encounter reflecting surfaces at different distances from the source, the two sets of waves will end up somewhat out of step, because one has traveled a greater distance than the other. To the extent that the trough of one wave encroaches upon the crest of the other the waves interfere destructively, and the reflected light is dimmed. At various angles of view the apparent

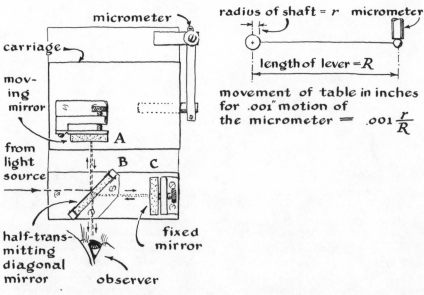

movement of table in inches for .001" motion of the micrometer $= .001 \frac{r}{R}$

*Path of light in an interferometer made by Eric F. Cave*

distance between the reflecting surfaces will be greater or less, and the intensity of the reflected light will appear proportionately brighter or dimmer, as the case may be. The total energy of the incident light remains unchanged by the interference. It is only the angles at which the energy is reflected that change. Hence the positions of the fringes with respect to the reflecting surfaces appear to shift when the observer moves his head. Similarly, the apparent position of the fringes depends on the length of the waves. Long waves of red light may appear to annul one another completely in a certain zone, while the short waves of blue light may appear reinforced. In that case the zone will appear blue, although the light source may be emitting a mixture of both long and short waves. If the source is white light, a blend of many wavelengths, some of the colors are annulled and others are strengthened at a given angle of view, with the result that the fringes take on rainbow hues.

Similarly changes in the distance between the reflecting surfaces cause the fringes to shift, just as though the position of the eye had changed. That is why the fringes move when enough pressure is applied to bend two sheets of glass not in perfect contact.

Another interesting simple experiment is to place an extremely flat piece of glass on another flat piece, separating the two at one edge with a narrow strip of paper, so that a thin wedge of air is formed between them. When the arrangement is viewed under yellow light, the interference fringes appear as straight bands of yellow separated by dark bands which cross the plates parallel to the edges in contact. The number of yellow fringes observed is equal to half the number of wavelengths by which the plates are separated at the base of the wedge. When the paper strip is removed and the plates are brought together slowly at the base, the fringes drift down the wedge and disappear at the base, the remaining fringes growing proportionately wider. By selecting relatively large plates for the experiment, it is possible to produce a fringe movement of several inches for each change of one wavelength at the base. A version of the interferometer is based on this principle.

In short, any change which modifies the relative lengths of the paths taken by two interfering rays causes the position of the resulting fringes to shift. A change in the speed of either ray has the same effect, because the slowed ray will arrive at a distant point later than the faster one, just as if it had followed a longer path. Any material medium will

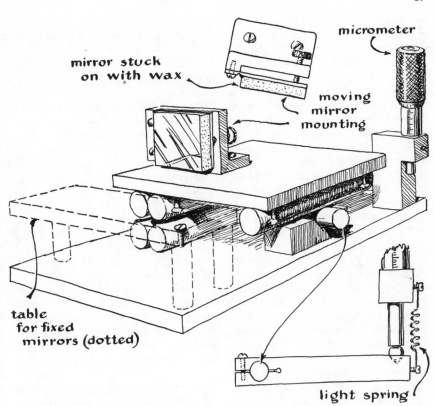

*Details of the interferometer base assembly*

slow light to less than its speed in a perfect vacuum. Air at sea level cuts its speed by about 55 miles per second, short waves being slowed somewhat more than long ones. If two interfering rays are traveling in separate evacuated vessels and air is admitted into one of the vessels, the interference fringes will shift, as if that path had been lengthened. From the movement of the fringes it is possible to determine by simple arithmetic the amount by which the speed was reduced. The ratio of the velocity of light through a vacuum to its velocity through a transparent substance is called the refractive index of that substance. The interferometer is a convenient instrument for measuring the refractive indices of gases and of liquids.

Eric F. Cave, a physicist at the University of Missouri, has designed a simple interferometer which will demonstrate many of these interesting effects and enable even beginners in optics to measure the wavelength of light. With suitable modifications the instrument can be used for constructing primary standards of length, measuring indices of refraction, determining coefficients of expansion and so on.

"The design presented here," writes Cave, "is intended to serve primarily as a guide. Most amateurs will be capable of designing their own instruments once the basic principles are understood. Optically the arrangement is similar to that devised by Michelson. A source of light, preferably of a single color, falls on a plate of glass which stands on edge and at an angle of 45 degrees with respect to the source. This plate serves as a beam-splitter. Part of the light from the source passes through the plate. This portion proceeds to a fixed mirror a few inches away, where it is reflected back to the diagonal plate. The other part of the original light beam is reflected from the surface of the diagonal at a right angle with respect to the source. It travels to a movable mirror, located the same distance from the diagonal plate as the fixed mirror, and it too is reflected back to the plate. Part of this ray passes through the plate to the eye. Here it is joined by part of the ray returned by the fixed mirror [see drawing on page 66].

"By adjusting the positions and angles of the two mirrors relative to the diagonal plate it is possible to create the illusion that the fixed mirror occupies the plane of the movable mirror. Similarly, by adjusting the angle of either mirror slightly, it is possible to create the optical effect of a thin wedge between the two mirrors. Interference fringes will then appear, as if the two reflecting surfaces were in physical contact at one point and spaced slightly apart at another. A change in the position of the movable mirror toward or away from the beam-splitter is observed as a greatly amplified movement of the fringes.

"Beginners may expect to spend a lot of time in coaxing the instrument into adjustment. But careful construction will minimize the difficulty.

"The base can be made of almost any metal, although amateurs without access to shop facilities are advised to procure a piece of cold-rolled steel cut to specified dimensions. The instrument can be made in any convenient size. The base of mine is nine inches wide and 14 inches long. You will also need two other plates of the same thickness and width but only about a quarter as long. They become the carriage for supporting the movable mirror and the table on which the diagonal plate and fixed mirror are mounted.

"The carriage moves on ways consisting of dowel pins attached to its underside [*see drawing on page 67*]. The ways are made of commercially ground drill rod, which can be procured in various sizes from hardware supply houses. Each way consists of a pair of rods, one set being attached to the carriage and the other to the base. The ways can be fastened in a variety of arrangements. I fitted them into a milled slot. Flat-headed machine screws will serve equally well as fastenings if you do not have a milling machine. The bearing for the

drive shaft can be a block cut with a V-shaped notch. If no shop facilities are available for machining it, you can drill four shallow holes in the base as retainers for four steel balls and simply let the shaft turn between the two sets. The height of the block or ball supports should be chosen so that the top of the carriage will parallel the top of the base when the machine is assembled. The ways move on two steel balls fitted with a ball-spacer made of thin aluminum as shown. In operation the carriage is driven back and forth by turning the drive shaft.

"The shaft may be rotated either by a worm and wheel arrangement or by a 'tangent screw.' The latter consists of a screw pressing on a bearing in a lever arm, the other end of which is attached to the shaft [*see detail at lower right in drawing on page 67*]. A tangent screw permits only a small amount of continuous travel, but it is less expensive than a worm and wheel.

"The lever arm should be rectangular in cross section. One end is drilled for the shaft, split as shown and fitted with a clamping screw. The other end is drilled with a shallow hole for the steel ball-bearing. The screw may be a ma-

chinist's micrometer mounted on a bracket as shown. The ball bearing is held in close contact with the micrometer by a spring. The length of the lever and the diameter of the drive shaft determine the amount by which the carriage will move when the micrometer is turned.

"It should be possible to control the movement of the table smoothly through distances equal to at least one wavelength of the light under investigation. The wavelength of the yellow light emitted by glowing sodium is about one 50,000th of an inch. The tangent screw must therefore provide a geometrical reduction to distances of this order. When the machinist's micrometer is turned one division, the screw moves the outer end of the lever arm a thousandth of an inch. By adjusting the effective length of the lever arm (the distance between the center of the ball bearing under the screw and the center of the shaft) with respect to the radius of the shaft, the relative movement of the carriage can be reduced by any proportion desired. The reduction is equal to the radius of the shaft divided by the effective length of the lever arm. Thus a 10-inch arm coupled to a 1/4-inch shaft yields a reduction of 80 to 1, and a turn of one micrometer division produces a carriage translation of .0000125 of an inch.

"The quality of the optical parts will largely determine the experiments possible with the instrument and the extent to which it may be worth while later to add accessories and otherwise modify the design. Advanced telescope makers will doubtless prefer to grind and figure the three flats required. Those less skilled in figuring glass may order them from an optical supply house. All three elements should be flat to about a tenth of a wavelength or the resulting fringes will show serious distortion. Small squares can be cut from plate glass and tested for flatness by the method outlined in *Amateur Telescope Making* by Albert G. Ingalls. If the instrument is to be used for testing lenses, mirrors, prisms and so on, the faces of the pieces of glass should be strictly parallel to one another. Both the fixed and the movable mirror should be silvered or aluminized on the front surface, and for best results the face of the diagonal plate also should be silvered slightly, so that it will reflect about as much light as it lets through.

"Mounting brackets should support the optical elements perpendicular to the plane of the base after assembly. They should provide for finely controlled angular adjustment of the mirrors around the horizontal and the vertical

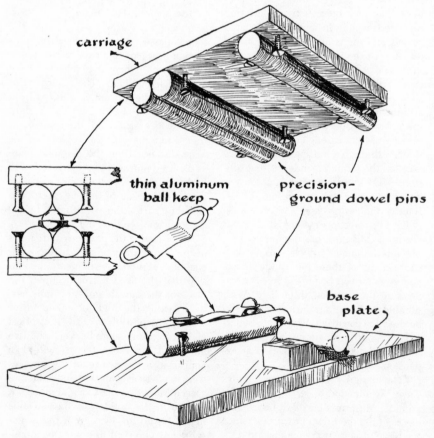

*Details of the interferometer carriage assembly*

axes. In the illustration here [*at right*] the movable mirror is mounted with wax, but for anything more than an initial demonstration this is not good practice, especially if the supporting member is subject to flexure. The diagonal plate and fixed mirror are mounted on a rectangular table fixed to the base, and are located so that the center of the beam of light from the source strikes the center of the diagonal plate and is reflected at right angles to the center of the movable mirror.

"Two important conditions must be fulfilled if the instrument is to function properly. The light must originate from an extended source several feet away, and it must be monochromatic. The yellow flame obtained with soda glass is not strictly monochromatic, because most of the light comes from the brilliant spectral doublet of sodium, but it is adequate for demonstrating the instrument.

"When in operation the instrument should rest on a solid, vibrationless support. The movable mirror is placed as precisely as possible at the same distance from the beam-splitter as the fixed mirror. Preliminary adjustments are then made with the aid of a point source of light—*e.g.*, the highlight reflected from a small polished steel ball 1/16 of an inch in diameter, placed about 10 feet away. The ball should be lighted with a concentrated beam such as that provided by a 300-watt slide projector. The ball is located to the left of the observer when he faces the movable mirror and in line with the center of the beam-splitter and fixed mirror. When you look at the movable mirror through the beam-splitter, you will see two images of the source. You change the angles of both mirrors by means of the adjusting screws until the images of the source coincide. Now you substitute the sodium source of light for the ball. If the distances of the mirrors from the beam-splitter are essentially equal, you should see a number of concentric circles in orange and black like those of a rifle target. The orange color is characteristic of the sodium doublet, while the black circles mark zones of destructive interference between beams reflected by the two mirrors. Remember that you are making exquisite adjustments requiring patience.

"To measure the wavelength of sodium light, first note the precise position of the micrometer and then turn it slowly

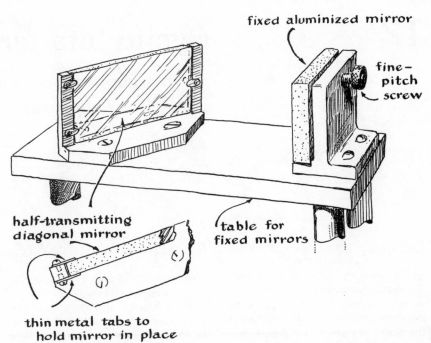

while counting the number of times the bull's-eye of the target changes from orange to black and back to orange again. From orange to orange or black to black is a half wavelength. Count, say, 100 of these color changes. The carriage has then moved 100 times the half wavelength of sodium light. Read the micrometer setting and subtract it from the first reading. This difference, when divided by the geometrical reduction provided by the tangent screw, is equal to 50 wavelengths of the light. In reality we are working with the sodium doublet, of course. The wavelength of one of the sodium lines is .000023188 of an inch and the other is .000023216 of an inch. If your experiment comes out correctly, therefore, your instrument will show the average of the two, or .000023202 of an inch. The fact that you are dealing with light of two close wavelengths may cause poor contrast between the orange and dark fringes at certain positions of the carriage. A slight displacement of the carriage from this position will produce maximum contrast.

"This design is intended merely to whet an appetite for interferometry. The instrument and theory discussed here are mere introductions to the subject. Before the instrument can yield results comparable with those achieved by Michelson, it must be provided with a mono-

chromatic source, such as the light emitted by the red line of cadmium. Advanced instruments are provided with a small telescope for viewing the fringes. In addition, Michelson inserted a second diagonal plate in the path between the beam-splitter and the movable mirror. It is unsilvered but otherwise identical with the beam-splitter. This plate equalizes the thickness of glass traversed by the two beams and prevents the short waves in one beam from being retarded more than those in the other.

"Interferometers can be equipped with accessories for measuring the physical constants of solids, liquids and gases. Glass containers provided with optically flat windows can be introduced into the beams. When the air in one is slowly displaced with a gas, the fringes will drift across the field just as if the carriage were being moved. A count can easily be reduced to the index of refraction of the gas. The coefficient of expansion of a solid with respect to changes in temperature can be determined by clamping the specimen, fitted with a thermometer, between the carriage and base. A count of the fringes is then converted into the dimensional change of the specimen. This information, together with the temperature difference, enables an experimenter to compute the desired coefficient."

*Details of the interferometer beam-splitter and fixed-mirror assembly*

# 12 Cyclic Interferometer

*An interferometer constructed from
plate glass and lenses*

*February 1973*

In a few of its several forms the interferometer serves as a convenient instrument for observing and recording photographically local variations in the density of a gas, such as disturbances caused in air by the heat of an open flame or the flow of air around an object in a small wind tunnel. Essentially the instrument splits a source of light into two beams of parallel rays, recombines the beams at a distant point and focuses the converging rays on a screen. The optical parts of the instrument, which consist of mirrors and lenses, can be adjusted so that light waves in one beam of parallel rays fall slightly out of step with those in the other beam. When the beams are recombined and projected on the screen, the crests of the waves will coincide in some areas of the screen to form bright patterns. In other areas the crests of the waves in one beam will coincide with the troughs of the waves in the other beam; the screen appears dark in these areas.

When the instrument is appropriately adjusted, the screen displays a gridlike pattern, being crossed by fringes, or alternately dark and light bands, that are uniformly spaced and parallel. If either of the split beams traverses a gas of nonuniform density in an instrument so adjusted, the form of the grid is altered to a pattern that corresponds to the density variations. Interferometers of conventional design must be made with parts of good optical quality because the projected image displays imperfections in the lenses and mirrors as well as nonuniformities in the density of a gas. Instruments of this quality are priced beyond the reach of most casual experimenters. Chris F. Bathurst of Invercargill in New Zealand has devised a scheme for making a useful interferometer from scraps of ordinary plate glass and with lenses of the kind used in inexpensive magnifying glasses. He discusses the construction as follows:

"My instrument is a variation of the cyclic interferometer, so called because the test and reference beams of light traverse a triangular path in opposite directions. The unusual feature of the instrument is a lens that is inserted in the base of the triangular path. The lens can be adjusted to minimize the effect of imperfections in the optical parts.

"Rays of light from a conventional source are focused into a converging cone that falls on a beam splitter in the form of a partially silvered mirror [*see illustration on opposite page*]. The apex of the transmitted cone passes through the test section of the instrument—the vessel that contains the gas specimen. The cone is deliberately located in a region of the vessel that is known to contain stable gas.

"The rays proceed through the test section as a diverging cone and are reflected by a mirror to the lens in the base of the triangular path, as shown by the colored rays in the illustration. The lens bends the rays into a parallel bundle that proceeds to the second mirror, which reflects the bundle to the beam splitter, thus closing the triangular path. Half of the light is reflected to the source by the beam splitter and is lost. The beam splitter transmits the remaining half to a field lens that focuses the rays at the viewing position. Essentially these rays function as a reference beam with which rays that are disturbed in the test chamber are compared.

"Converging rays from the source that are reflected by the beam splitter traverse the triangular path in the opposite direction, as shown by the black rays in the illustration. Following this reflection they impinge on the second mirror and are reflected to the lens in the base of the triangular path. The lens transforms the diverging cone into a beam of parallel rays. After subsequent reflection by the first mirror the beam floods the full area of the test section with parallel rays. The direction and the velocity of the rays will be altered more or less by local variations of density in the gas.

"Rays that emerge from the test section return to the beam splitter. Half of the light proceeds through the beam splitter to the source and is lost. The remaining rays are reflected by the beam splitter, interfere with rays of the reference beam and proceed through the third lens to the focal plane at the viewing position. If the gas in the test section is of uniform density, the instrument can be adjusted to project a pattern of fringes that are reasonably parallel and straight even though the surfaces of the optical elements are imperfect. The effects of optical imperfections can be reduced about 80 percent. The residual imperfections in the fringe pattern of the instrument described here are about seven fringes in a field of view three inches square.

"None of the lenses I used is of high quality. All are symmetrical. Lens 1 (focal length 10 inches, diameter four inches) is from a magnifying glass. Lens 2 (focal length 42 inches, diameter four inches) and lens 3 (focal length 25 inches, diameter four inches) were made by a local craftsman with a machine that is normally used for grinding and polishing spectacle lenses. Lens 4 (focal length 5½ inches, diameter 1¼ inches) is of relatively small diameter and reasonably good optical quality. Most of the light passes through it close to its axis. Hence the lens introduces little distortion.

"The green filter (Wratten No. 64) restricts the light in the image to monochrome. I use it for isolating the 5,461-angstrom line emitted by a mercury lamp. The filter can be omitted if a laser is used as the light source.

"All mirrors and the windows of the test section were made of ordinary plate glass 3/8 inch thick. The reflecting surfaces are films of aluminum deposited in a vacuum. A series of partially coated beam splitters was made to determine the optimum thickness of the aluminum in terms of reflection. I found that the thickness of the film is not crucial. The metallized coating must be on the surface of the beam splitter that is nearest the test section.

"The mirrors are supported at three points by mechanical mounts that were improvised from available materials. Small pieces of cardboard separate the metal clamps from the glass. The windows of the test chamber are sealed with gaskets of cork. Two micrometer screws, which function as adjustments, rotate the mirror mounts about the horizontal and vertical axes in the plane of the mirror.

"In addition to these adjustments the mounting of lens 2 must be adjustable in both directions parallel to the optical axis and at a right angle to the axis. I did not fit this mounting with screws for making translational adjustments, although they would be convenient. The base of the instrument is made of 1/4-inch steel channel sections and measures three inches in width by 1½ inches in height. Photographs of fringe patterns were made with a 35-millimeter camera.

"The adjustment procedure is somewhat tedious, as it is with most interferometers. My instrument can be aligned and adjusted in from 10 to 30 minutes, depending on the experience of the operator. After the mountings of the optical elements have been assembled to the base remove the beam splitter and lenses 2 and 3. Put a rectangle of white cardboard at the position normally occupied by lens 3. Move lens 1 toward the light source to the point at which the lens projects a light beam of substantially constant width. Align the mirrors to the position at which the beam illuminates the screen.

"Adjust lens 1 and mirrors 1 and 2 to focus a sharp image of the light source on the center of the screen. Replace the beam splitter. Two images of the source should appear on the screen. Adjust the beam splitter to the position at which the two images coincide exactly.

"Move lens 1 toward the light source. A pattern of light and dark fringes should appear on the screen. If fringes do not appear, return to the previous step and again attempt to adjust the beam splitter to the exact point at which

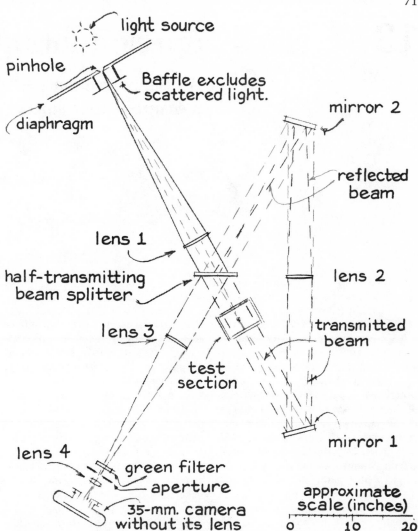

*Optical path of Chris F. Bathurst's cyclic interferometer*

the images of the source coincide.

"Look for the fringes. Repeat the procedure until they appear. The operation calls for patience until you acquire experience. After the fringes appear adjust the beam splitter to the position at which you observe the minimum number of fringes.

"Replace lens 2 and move lens 1 toward the beam splitter to the point at which the apex of the cone of converging light rays lies inside the test section. The location of the apex within the test section can be altered by shifting lens 1 up and down or sideways. Avoid regions where the apex would interfere with the phenomenon under study.

"Next, move lens 2 along its axis to the position at which fringes fill the field of view. If the fringes do not appear, return to the critical step in which the images of the light source were made to coincide by manipulating the beam splitter. When the field of view eventually fills with fringes, manipulate lens 2 and the mirror mounts to remove as many fringes as possible. Adjust ei-

ther of the mirrors to the position at which a desired pattern of fringes, such as a grid, appears in the field of view.

"Finally, transfer the image to the focal plane of the instrument by replacing lens 3, the field lens. Adjust the position of aperture near the camera to intercept the halo of stray light that surrounds the converging rays. The position of the focal plane of lens 4 can be located with a screen of white cardboard.

"To make sharp photographs of fringes that arise from phenomena in the test section place the film of the camera in the focal plane of the instrument. I have used light sources of three kinds. A frosted incandescent lamp can be located behind lens 1 for adjusting the camera to the position of sharpest focus with respect to phenomena in the test section. All other adjustments have been made with a 250-watt, high-pressure mercury arc lamp. Photographic exposures are made by replacing the mercury lamp with a xenon flash lamp of the kind available from dealers in photographic supplies."

# 13 Speckle Interferometer

*A laser interferometer that can measure displacement*

February 1972

When a moving object is photographed in ordinary light by time exposure, the resulting image is more or less blurred. Usually such photographs are of little interest to experimenters even though the fuzziness of the image varies with the amplitude of the movement. A comparable photograph made with coherent light, however, yields an accurate measurement of the displacement of the object. For example, a photograph of a vibrating tuning fork that is made with the coherent light of a helium-neon laser can be used for determining to less than a thousandth of an inch the amplitude of movement of the tines.

Displacements can be measured by making a double exposure of an object on a single sheet of film. One exposure is made before the object moves and one is made after it moves. The technique, which is known as speckle interferometry, can also be used to map local deformations in stressed mechanical parts such as the components of telescope mountings, seismometers, optical benches and similar devices. Ulrich Köpf of Fleischmannstrasse 7 in Munich explains the procedure as follows:

"Speckle interferometry resembles conventional holography in some respects but is much simpler to do. LeRoy D. Dickson has presented in these columns several ingenious hints for reducing the sensitivity of holographic apparatus to vibrations [see "The Amateur Scientist"; SCIENTIFIC AMERICAN, July, 1971]. Nevertheless, good holograms are not easy to make. In contrast, a reasonably careful beginner who undertakes an experiment with speckle interferometry can expect success on the first try.

"The technique is based on the speckling or granularity that appears when laser light is reflected by a dull surface such as a painted wall, a sheet of white paper or a roughened piece of metal. The uniform illumination produced by an incoherent source, such as an incandescent lamp, is replaced by a pattern of dazzling granules when a coherent source is used. Each speckle marks a point of constructive interference between waves of coherent light. The speckles are of high contrast. The diameters of the speckles vary statistically, but the mean diameter is determined by the resolving power of the optical system with which the speckles are observed and is approximately equal to the product of the aperture of the pupil of the eye (in the case of a camera, the focal ratio, or $f$ number) multiplied by the wavelength of the light. The wavelength of the coherent light emitted by a helium-neon laser is $6.33 \times 10^{-7}$ meter (6,330 angstroms).

"The pattern of speckles is caused by the diffuse surface texture of the object. When the surface is displaced at a right angle to the direction from which it is viewed, the pattern of speckles moves with it, much as if the speckles were minute spots of paint. The pattern can be recorded on photographic film as a double image. The amplitude of the displacement can be determined by examining the pattern of speckles.

"With these principles in mind, one begins the experiment by flooding an object with the expanded beam of a helium-neon laser. The object might be a simple metal clamp similar to the one shown in the accompanying illustration [*top of page 108*]. The object can be supported by any reasonably stationary base, such as a brick. The beam of laser light can be expanded into a cone with a lens that has a focal length of some four to eight millimeters. I use a 40-power objective lens from a microscope, but a simple plano-convex lens will work as well. The distance between the laser and the object is not crucial, but it should be adjusted so that the cone of coherent light floods all parts of the object. The lens can be held in position by any improvised support that is reasonably rigid. The camera is placed at a distance where the sharply focused image of the object occupies a substantial portion of the photographic negative. The negatives should be of the high-resolution type. I use four-by-five-inch Agfa Gevaert Scientia plates, which are similar to Eastman Kodak 649F plates.

"My exposures are made with a studio camera. After the camera has been focused the aperture is set at a focal ratio, or $f$ stop, that varies according to the resolving power of the photographic emulsion and is equal to 1 divided by the product of the resolving power of the emulsion multiplied by the wavelength of the coherent light. (The resolving power of the emulsion is specified by the manufacturer.)

"The exposure interval is determined by flooding the object with laser light and measuring the intensity with a commercial exposure meter. Make the first exposure. Displace the object or otherwise alter its position in whole or in part at a right angle to the direction of the camera. With an object such as the clamp used in my experiment the metal can be warped by heating one side of the device with a small torch. The second exposure is then made on the same negative. The negative records two displaced but otherwise identical patterns of speckles. The negative is developed and fixed according to the procedure recommended by the manufacturer.

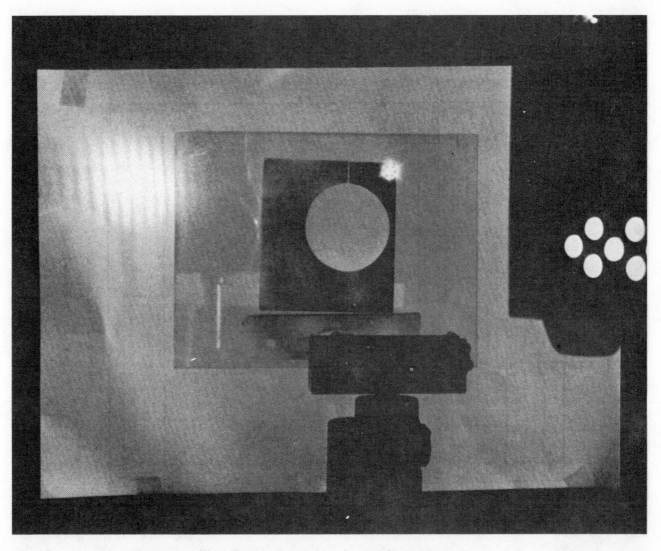

*Ulrich Köpf's apparatus for making speckle interferograms*

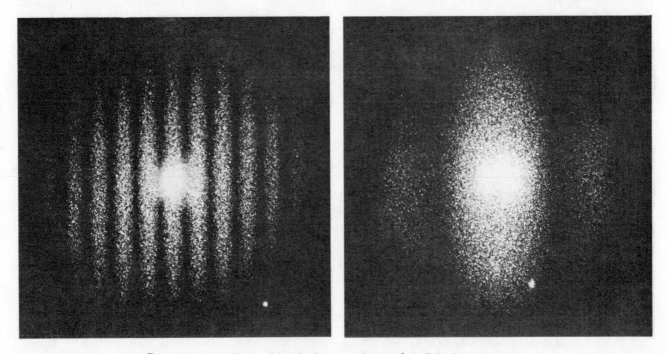

*Fringe patterns indicating large displacement* (**left**) *and small displacement* (right).

laser

beam-expanding
microscope objective

image of object

object

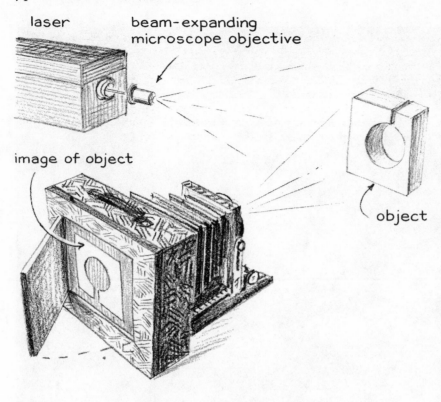

*Equipment for Köpf's experiments*

laser

focal length
of lens

negative

transformation
lens

screen

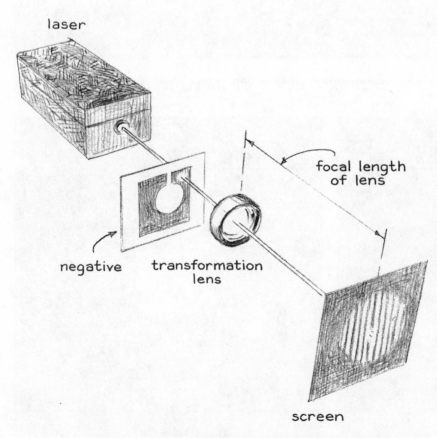

*Arrangement for examining the holographic interferogram*

"The developed plate resembles an ordinary photographic negative, but the exposed area shows speckles. To detect the displacement and measure its amplitude the experimenter must analyze the complex speckle pattern into its local components by means of a Fourier transform. Do not be dismayed by the prospect of having to carry out this mathematical procedure with paper and pencil. It can be accomplished by an inexpensive analogue computer: a simple lens.

"To make the analysis support the film in the vertical plane by an improvised clamp and direct the laser beam through part of the image. Intercept the transmitted beam by the transformation lens adjusted to focus an image on a distant screen [*see bottom illustration at left*]. The image consists of parallel fringes of light and shade. The spacing of the fringes varies inversely with the displacement of the object at the point on the negative through which the laser beam is transmitted.

"Measure the distance $d$ between any pair of adjacent fringes. It is easy to calculate from this known distance the displacement $s$ of the object. The calculation includes the wavelength of the laser light ($l$), the focal length of the transformation lens ($f$) and the magnification of the photographic image ($m$). Usually the image is smaller than the object, hence the magnification is represented by a fraction. To determine the magnification divide the width of the image, as measured with a ruler, by the width of the object. The amplitude of the displacement is calculated by means of the formula $s = l \times f / m \times d$.

"In my experiment with the metal clamp the laser beam was directed through the image at two points: the upper corner, where maximum displacement was expected, and at the thin edge on one side [*see illustrations on preceding page*]. Fringes associated with the upper corner of the part were spaced $6 \times 10^{-3}$ meter apart. Fringes associated with the thin edge measured $3.8 \times 10^{-2}$ meter. The focal length of my transformation lens was $3 \times 10^{-1}$ meter and the magnification of the image was .7. The displacement of the upper right corner of the clamp was therefore $6.33 \times 10^{-7} \times 3 \times 10^{-1} / .7 \times 6 \times 10^{-3} = 46 \times 10^{-6}$ meter, or .0018 inch. The displacement of the edge of the clamp was similarly calculated: $6.33 \times 10^{-7} \times 3 \times 10^{-1} / .7 \times 3.8 \times 10^{-2} = 7 \times 10^{-6}$ meter, or .00028 inch.

"The method is particularly easy to apply in the case of vibrating objects such as a tuning fork because the motion can be photographed by a single exposure. The amplitude of vibration can be measured throughout the length of the vibrating part by directing the laser beam through the photographic negative at a series of points. The laser need not be adjusted for operation in a single mode, as in making holograms, but can oscillate simultaneously in many transverse modes. Local deformations in an object that is stressed but not moved between two exposures are measured by scanning the photographic negative with the laser beam and monitoring the changing fringe distances and directions.

"Displacements occur at right angles to the direction of the fringes. Incidentally, if the transformation lens is removed, the fringes can still be observed provided that the diameter of the beam is less than the fringe separation."

A warning: Laser light is hazardous, particularly to the eye. *Never look into the beam.* When you are using a laser, confine the beam and all targets within an opaque housing.

# 14 Series Interferometer

*A series interferometer to observe*
*various subtle phenomena*

June 1964

For the precise determination of length, or of small changes in the density, pressure or temperature of transparent substances, amateur experimenters will find no yardstick more accurate than a simple beam of light as used in a homemade interferometer. All interferometers are based on the principle that light waves that take different paths from a common source can fall out of phase and either cancel or reinforce each other when they reunite. If the source consists of white light, which is a blend of many wavelengths, the interfering waves produce colorful patterns as seen in such natural interferometers as soap bubbles, opals and all colored bird feathers except a few with yellow pigmentation. Interferometers that are commonly used for measuring length split a beam of monochromatic light into two rays that take separate

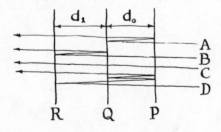

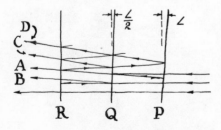

*Paths of rays through partial mirrors*

paths until they recombine to interfere. The resulting pattern appears in a monochrome of varying intensity, as discussed in this department some time ago [November, 1956]. Instruments of the type used for measuring the density of fluids or gases, as well as the local distribution of temperature and stress or pressure, also split the beam into two rays, but the rays take essentially the same path. These instruments, known as series interferometers, cause one ray to traverse part of the path more than once; interference occurs at the end of the transit.

The construction of most series interferometers requires the use of machine tools not ordinarily available to amateurs. However, a series interferometer that can be made with ordinary hand tools has now been designed by G. F. Pearce, professor of mechanical engineering at the University of Waterloo in Canada. "The essential part of this instrument," writes Pearce, "consists of a series of three partially transparent mirrors. A beam of light, when passing through the mirrors, may take many paths of differing length. A given ray, for example, may pass through the first mirror, be reflected from the second back to the first and then back again through both the second and the third mirror. A second ray that substantially coincides with the first may traverse both the first and the second mirror and be reflected from the third back to the second for a final reflection to and through the third. Still another ray may traverse the first two mirrors, be reflected from the third back to the first, oscillate for a time between the first and the second mirror and finally complete its transit through the third, as shown in the upper part of the accompanying illustration [this page].

"When the mirrors are about equally spaced and are approximately parallel, the rays that traverse paths A and B, as depicted in the illustration, will interfere to form a pattern of alternate light and dark areas. Other parts of the beam that traverse a variety of path lengths, such as those shown at C and D, form an interference pattern that is superimposed on the interference pattern caused by A and B. This superimposed pattern can be eliminated by inclining the mirrors as shown in the lower part of the illustration. If mirrors P and Q are inclined as indicated, the rays that traverse different paths become separated and emerge in different directions.

"The interference pattern can be observed by simply inserting a collimator lens between the source and the mirror system so that the light passes through the mirrors as a bundle of parallel rays that are focused on a screen by a field lens between the third mirror and the screen [see top illustration on opposite page].

"The mechanism I contrived for supporting the mirrors in any relative position was constructed of aluminum plates. Any other substantial material, such as plastic panels or even plywood, may be substituted, however. The partial mirrors can be made by the familiar techniques used by amateurs for silvering the mirrors of reflecting telescopes or can be obtained from suppliers such as the Edmund Scientific Co. in Barrington, N.J., or Henry Prescott in Northfield, Mass. The mirrors I use reflect approximately 40 per cent of the incident light and transmit 60 per cent. They measure about two inches wide by three inches long, but the size is not critical. Homemade instruments of the same type have been operated with partially silvered microscope slides that measure one by three inches.

"The base of the interferometer should be constructed of material at least three-quarters of an inch thick to provide a solid support for two equally sturdy end brackets [*see middle illustration on this page*]. Apertures that are just a fraction of an inch smaller than the mirrors are cut in the end brackets. One mirror is mounted over the aperture of the bracket that faces the screen. The glass can be attached to the metal by light clips of spring brass or by a few dabs of epoxy cement. The reflecting surface should face away from the screen. The remaining two mirrors are similarly attached over the apertures of two intermediate supporting plates, which can be made of thinner material. One of the intermediate plates is attached to the bracket nearest the screen by a suspension system that consists of four adjusting screws, together with a set of four helical compression springs that space the plate from the bracket. The bracket is drilled and tapped in its four corners for the screws, but the plate is merely drilled with oversize holes that slide easily over the screws when the retaining nuts are turned. These nuts are normally used only for tilting the middle mirror in relation to the mirror closest to the screen. The middle mirror can be attached to its supporting plate by the same technique used for mounting the first mirror; the coated side should face away from the screen.

"The system used for supporting the third mirror is similar. In this case, however, the supporting plate is suspended from the second bracket by only three screws. In my instrument these screws are equipped with heads in the form of worm gears taken from surplus apparatus. Worm gears are not essential, but they are exceptionally convenient for making fine adjustments. Oversize holes were made for the screws in the end bracket; the plate that carries the mirror was drilled and tapped. The threads make a loose fit, so the screws do not bind when the plate is tilted a few degrees. As shown in the accompanying illustration [*bottom of this page*], compression springs maintain the desired spacing between the end bracket and the plate. The coated surface of the third mirror should face toward the screen. The completed interferometer can be supported on any substantial surface, such as the top of a solid bench or table.

"The lenses used for bending the light into a bundle of parallel rays that traverse the mirrors and for focusing

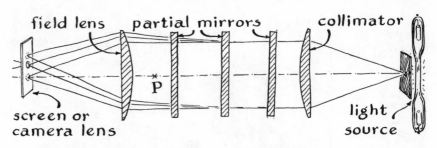

*Optical scheme of a series interferometer*

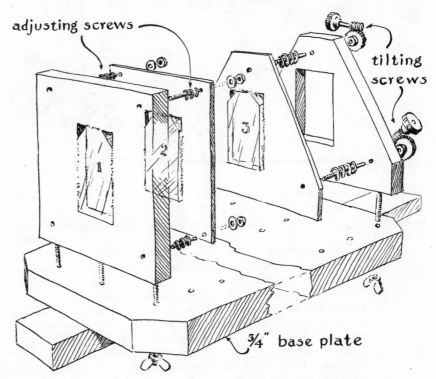

*Arrangement of mirrors and brackets*

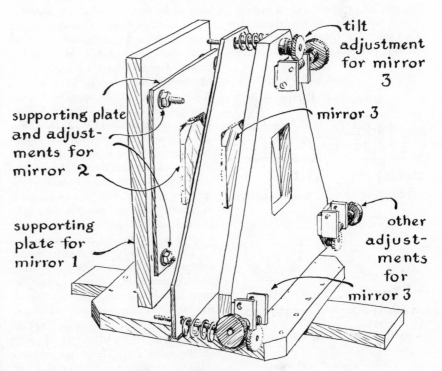

*Details of the mirror-suspension system*

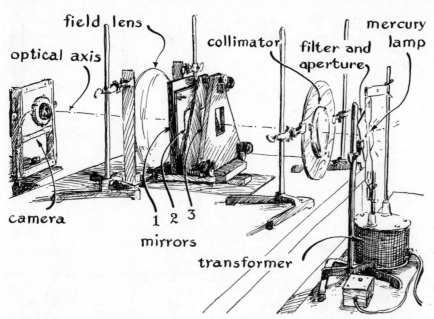

*A series interferometer designed for construction by an amateur*

it on the screen are of the simple plano-convex type. They need be no larger in diameter than the mirrors. Neither do they have to be achromatic, because the light is monochromatic. My lenses were obtained from the Edmund Scientific Co. A bracket for supporting the collimating lens was improvised from a piece of plywood that in turn is held in position by an apparatus stand. The position of the lens is adjusted simply by moving the clamp and stand. A comparable arrangement supports the field lens between the mirror system and the screen. Although the interference pattern can be projected onto a screen, I prefer to examine it on the ground glass of a camera so that photographs of interesting patterns can be made conven-iently. I use a view camera equipped with an extension bellows. Cameras of this type, incidentally, provide many desirable features not found in minia-ture cameras of more recent design. Often an older model made of wood can be bought for less than $10, complete with a set of plateholders.

"As the source of monochromatic light I use a tubular mercury lamp to-gether with a green filter similar to the Corning Type 4-64 that blocks the transmission of all rays except those emitted by the 5,460-angstrom line of mercury. Rays from the lamp are re-stricted by an aperture three-sixteenths of an inch in diameter located at the focus of the collimating lens. An alter-nate source, somewhat less intense, could be provided by a mercury lamp of the General Electric Type H-100-A38-4, which must be operated in con-junction with a current-limiting ballast, such as a Type 9T64Y3518. Doubtless an adequate source could be impro-vised by placing a sheet of dark green plastic, of the kind used for candy wrappers, over the mercury bulb of an ordinary sunlamp, but the precise color of plastic and the number of thick-nesses to use for optimum results would have to be determined experimentally.

"The fully assembled interferometer occupies a space about two feet wide and four feet long. Usually I mount the lamp on a small stand on one side of the table that supports the mirror as-sembly and lenses. The camera rests on a tripod just beyond the opposite side of the table [see illustration above]. With the apparatus so positioned, I first

adjust the mirrors for approximately equal spacing; in a typical experiment that might be five-eighths of an inch. Next I check to ensure that the mirror facing the camera (mirror 1 in the illus-tration) is perpendicular to the base plate. The second mirror is then tilted away from the camera approximately one degree and the third mirror ap-proximately two degrees.

"Next, the system of mirrors is aligned to prevent multiple images from reaching the ground glass when the image of the source is focused by the lens of the camera. Normally this will require both twisting and tilting the interferometer [see bottom illustration at left]. As an aid in making the ad-justment I insert a disk of transparent plastic about an inch in diameter and three-sixteenths of an inch thick be-tween the two mirrors closest to the light source. The adjustments are made simply by slipping wedges of appro-priate thickness between the base and the table top either to incline the as-sembly toward the camera slightly or to rotate it about its longitudinal axis. If overlapping disks of light in vertical array are observed on the ground glass, the instrument must be tilted; if the disks overlap horizontally, twist is re-quired. When the position of the base has been altered so that the images merge, the orientation of the inter-ferometer is correct.

"During the next procedure—adjust-ing the mirror spacing—it is convenient to view the image of the light source

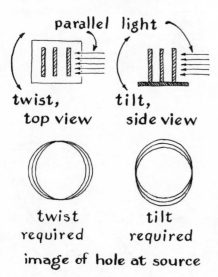

*Initial adjustment of images*

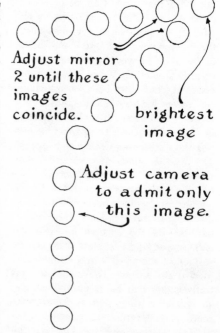

*Final adjustment of images*

from a position about midway between the field lens and the first mirror, that is, from position *P* in the accompanying illustration [*top of page 77*]. To view the image, the rays can be deflected to the side by inserting a small hand mirror between the field lens and the interferometer mirror, or by removing the field lens temporarily. The image will appear as a series of bright spots in echelon formation that diminish in intensity as shown in the upper part of the illustration [*bottom right p. 78*]. Adjust the center mirror by the screws that suspend it from the end bracket until the two images closest to the apex of the V coincide. Fine lines of light and shade—the interference fringes—should now be seen. The reason for a solid supporting table will now become apparent: the slightest vibration will disturb the fringes and make them difficult to see. Replace the field lens. A vertical array of images should now be visible at its focal point, as shown in the lower part of the illustration.

"Position the camera so that its iris coincides with the focal point of the field lens. Then alter the position to exclude all images from the ground glass except the second from the top, as illustrated. When the camera is properly focused, an interference pattern of alternate dark and light lines will be seen on the ground glass. The spacing of the lines is determined by the thickness of the plastic disk, as shown by the accompanying illustration [*at left below*]. If the intensity of the image is uncomfortably low, the pattern can be examined directly by removing the ground glass and observing the lines by means of a small magnifying lens, such as a jewelers' loupe. The fringes can be altered in inclination and spacing by adjusting the tilting mechanism of the bracket nearest the light source.

"When so adjusted, the interferometer is sensitive to the optical properties of any transparent substance placed between the mirrors closest to the light source. The transparent disk of plastic can be used for an initial test. If a mechanical load is applied across the diameter of the disk, for example, the resulting stress will reduce the diameter of the disk and increase its thickness. The distortion will not be uniform. Accordingly some of the transmitted light rays now traverse a path that is optically longer or shorter than other rays, as indicated by the altered interference pattern. To determine the nature of the distortion, first make a photographic negative of the interference pattern of the unstressed disk and a second negative with the disk loaded. Superimpose the negatives in register and make a photographic print through the pair. The print will show the kind of distortion pattern indicated in the accompanying illustration [*bottom right on this page*].

"The temperature distribution in air or other transparent media near a hot surface can be determined by observing changes in the density of the medium as reflected by the altered index of refraction. Heat lowers the density of most media and increases the velocity of light in the region of lower density, thus lowering the index of refraction. The effect can be demonstrated by placing a heated wire in front of the transparent plastic disk in the interferometer. Whereas the fringes appear to meet the image of a wire at room temperature at right angles, as depicted at the upper left in the accompanying illustration [*at right*], they curve increasingly as they approach the image in the case of a heated wire, as shown at the upper right in the illustration.

"Such patterns can be analyzed in terms of the temperature distribution. Assume the pattern of fringes associated with a heated wire, such as the one in the lower part of the illustration. Select a pair of adjacent fringes, such as those at *s* and *t*. A line drawn parallel to the straight portion of the fringes and continued toward the wire will intersect fringes that depart from straightness as they approach the wire. At point *v* in the illustration, for example, the straight, broken line that has its origin in the straight portion of fringe has crossed the bent-up portion of fringe *t*. The heated wire has caused a shift of one whole fringe with respect to the center line of fringe *s*. This means that the

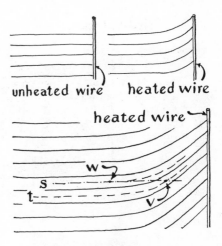

*Various fringes from wire*

light passing through point *v* has speeded up and is a full wavelength ahead of the light at point *w*. The amount of shift in terms of wavelength can be similarly determined at any point along the fringe. The index of refraction of a perfect vacuum has been accepted as 1, and that of air is about 1.0003. The density of normal room air can be measured by a barometer. With these data plus the number of wavelengths that the light shifts, as measured by the interferometer, the experimenter can determine the index of refraction of the heated medium and the local temperature by simple arithmetic, using formulas in the illustration [*p. 80*].

"The first of the equations, on which the formulas are based, was derived independently during the past century by the physicists H. A. Lorentz of Holland and L. V. Lorenz of Denmark. It relates the index of refraction of a gas to its density. As indicated, the index of refraction, −1, is very closely equal to the product of the density of the medium multiplied by a constant. The constant is therefore equal to the quo-

*A typical fringe pattern*

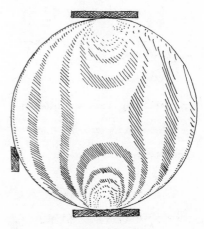

*Plastic disk under stress*

$$\frac{n^2-1}{n^2+1} = C\rho$$

where: $n$ = index of refraction

$\rho$ = density

$C$ = a constant

$$\frac{(n-1)(n+1)}{n^2+1} = C\rho$$

i.e., $(n-1) = C\rho$ very closely

$$n_1 - n_2 = C(\rho_1 - \rho_2)$$
$$= C\left(1 - \frac{\rho_2}{\rho_1}\right)$$
$$= (n_1-1)\left(1 - \frac{\rho_2}{\rho_1}\right) \ldots \text{①}$$

$$N_o = \frac{l}{\lambda_o}$$

where: $l$ = length of light path in heated air

$\lambda_o$ = wavelengh of light in vacuum

$N_o$ = number of wavelengths along $l$ in a vacuum

$$n_1 = \frac{N_1}{N_o} \text{ and } n_2 = \frac{N_2}{N_o}$$

where: $N_1$ and $N_2$ are numbers of wavelengths under experimental conditions

substituting in equation 1:

$$\frac{N_1}{N_o} - \frac{N_2}{N_o} = (n_1-1)\left(1 - \frac{\rho_2}{\rho_1}\right)$$

$$N_1 - N_2 = (n_1-1)N_o\left(1 - \frac{\rho_2}{\rho_1}\right)$$

$$N_1 - N_2 = (n_1-1)\frac{l}{\lambda_o}\left(1 - \frac{\rho_2}{\rho_1}\right)$$

tient of the difference between the index of refraction and unity divided by the density, a quantity that can be measured by a thermometer and a barometer. If $n_1$ is the index of refraction of the room air and $n_2$ is the index of refraction of the heated air, and if $p_1$ and $p_2$ are the corresponding densities, then, as indicated in the second set of equations, it can be shown that the difference between the speed of light through the room air and through heated air is equal to the product of the difference in the speed of light through air at room temperature and vacuum multiplied by 1 minus the ratio of the density of the heated to the unheated air. Similarly, the third set of equations relates the difference between the number of wavelengths that traverse air at room temperature and any selected point in the heated region, as measured by the interferometer, to the index of refraction of air at room temperature and the ratio of the density of hot to unheated air.

"Finally, by applying the law that relates the temperature of a gas to its density the formula is derived for computing the temperature of the heated air. To apply the formula the experimenter need only measure the temperature, $T_1$, of the room air and express the measurement in degrees Kelvin. Because the quantity $\Delta N$ is equal to the fringe displacement as measured in fringe widths by the interferometer, the formula can be used to calculate the temperature rise above room air at any point along the center line of the interference fringe. The fringe pattern in the vicinity of a horizontal wire takes the form of a series of concentric ovals, as shown in the upper part of the accompanying illustration [at right]. This pattern can be analyzed by the same technique; the temperature data can then be used for making a graph of the isothermals as shown in the lower part of the illustration."

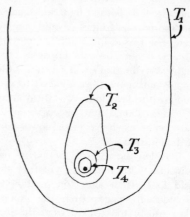

*Fringe (above) and isothermals (below)*

relating temperature to density:

$$N_1 - N = \Delta N = (n_1-1)\frac{l}{\lambda_o}\frac{(T_2-T_1)}{T_2}$$

solving for temperature:

$$T_2 - T_1 = T_1\frac{\Delta N}{(n_1-1)\frac{l}{\lambda_o} - N}$$

$\Delta N$ = fringe displacement measured in fringe widths

*Formulas and equations for interferometer*

# Interferometer to Measure Velocity

## A laser interferometer that converts a velocity to a sound signal

*December 1965*

I have recently repeated a fascinating experiment with the laser that was described in the *American Journal of Physics* for May, 1964, by David Dutton, M. Parker Givens and Robert Hopkins. The experiment suggests a number of applications for the apparatus, among them the detection of microscopic movements of an object at distances up to 100 feet. The beam is first collimated by a microscope objective with a focal length of approximately eight millimeters; it projects the rays to a lens of two-centimeter aperture and proportionately longer focal length. The collimated beam is then directed into a Michelson interferometer, the beam splitter of which can be an unsilvered microscope slide. One portion of the split beam proceeds a few centimeters to a plane front-surfaced mirror of optical quality; the other portion, which is transmitted at right angles by the beam splitter, is projected to a distant mirror of the cube-corner type.

Both the plane and the corner mirror are adjusted to return the reflected rays to the beam splitter, where they combine and interfere. The interference fringes are picked up by a photocell that acts as the input to an amplifier and a loudspeaker [*see bottom illustration on this page*]. When the corner mirror is moved at velocities of less than approximately one centimeter per second, the fringes modulate the photocell and a tone is emitted from the loudspeaker that varies in pitch in proportion to the velocity. The loudspeaker emits a rumbling sound even when the corner mirror is apparently at rest. This is explained by random vibrations, including microseisms, as well as by variations in the refractive index of the air. Numerous possible applications come to mind for the apparatus, including strain seismographs, the precise determination of length and the accurate monitoring of distant positions.

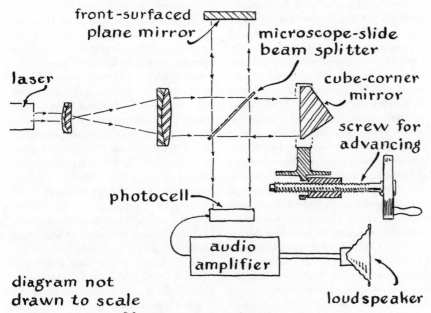

*Schematic arrangement of interferometer*

# 16 Interferometer to Measure Dirt Content of Water

*A laser beam and a photocell to measure the dirt content of water*

June 1973

The purity of the water in rivers, lakes and ponds can be tested by determining the coliform-bacteria count or the presence of dissolved gases and of various compounds. Tests of this kind were described in this department in March, 1970, and February, 1971. During the past year a remarkably accurate instrument for measuring turbidity (the concentration of solid matter that is carried by the water in the form of solid particles) has been developed by a group of undergraduates at the University of Rochester. The group, which included Cathie Lubell, Thomas Barry and Gregory Hearn, worked in the Institute of Optics and Materials Science Program under the supervision of Edward M. Brody, assistant professor of optics. They describe the work as follows:

"It is possible to determine the turbidity of solutions by many techniques. With a microscope one can count the number of particles in a unit volume of the specimen. With a calibrated grid in the viewing field of the microscope one can measure the relative sizes of the particles in a polydisperse suspension (a suspension containing particles of widely varying size). Turbidity can be evaluated relatively by filtering the specimen and weighing the dried filtrate. All these procedures are time-consuming. Boredom and fatigue on the part of the observer can lead to error.

"Extinction turbidimeters are doubtless the simplest instruments that have been devised for measuring the concentration of solids in suspension. They are based on the principle that turbidity is inversely proportional to the minimum length that a column of fluid must have in order to extinguish at one end of the column a source of light at the other end. A major source of error in all such turbidimeters arises from the fact that changes in the range of particle sizes in the suspension affect the pattern in which the transmitted light is reflected from particle to particle. The effect is known as multiple light-scattering. It causes apparent turbidity to decrease as the length of the light path through the specimen is increased. The influence of the multiple-scattering effect becomes increasingly significant when the attenuation of light through the specimen exceeds 15 percent. Extinction turbidimeters must therefore be calibrated with suspensions of known particle size. Indeed, reliable measurements of turbidity cannot be made with extinction turbidimeters unless the distribution of particle sizes is known.

"The substitution in turbidimeters of lamps for natural illumination and of photocells for the eye has led to improved measurements, but many sources of potential error remain. Variations in voltage applied to the lamps, fatigue of the photocathode of photocells and the effects of space charge in photomultipliers can combine to introduce an error of at least 10 percent. Moreover, it has been shown that the inherent temperature sensitivity of photoelectric detectors causes the output to vary as much as 50 percent through the temperature range from zero to 30 degrees Celsius.

"To cope with such errors we built an instrument of the ratio type in which an error in one part of the system is in effect compensated for, keeping the ratio constant. We call the device a dual-beam turbidimeter. Apparatus employing dual-beam transmission is available commercially, but it is costly and needs modification for use as a turbidimeter.

"In our device light from a single source is split into two beams. One beam traverses the specimen and the other, a reference beam, is transmitted through air. The two beams combine in a photodetector, where they are added algebraically. Experience has demonstrated that the instrument measures turbidity effectively whether the liquid contains suspended particles of widely varying size or suspended particles of uniform size. The results have been highly reproducible when checked against comparable measurements made by other procedures.

"A helium-neon laser serves as the light source [*see top illustration on opposite page*]. We elected to use the laser because it costs little more than a high-intensity lamp and provides parallel rays in the form of a monochromatic beam one millimeter in diameter. The output power of about two milliwatts exceeds the requirements of the photomultiplier detector by at least 100 times. We reduce the intensity by inserting neutral-density filters in the beams. An incandescent lamp could be substituted for the laser by inserting a color filter in the optical path (to select a range of wavelengths) together with simple collimating lenses to make the rays parallel.

"The output of the laser falls on a beam splitter, which is a rectangle of clear plate glass set at an angle of 45 degrees with respect to the beam. The transmitted portion of the beam passes through a neutral-density filter and enters the specimen cell. This vessel (a hollow prism made of sheet glass) has the form of a right triangle. The refracted rays emerge from the cell, pass through an aperture and fall on the photocathode of the photomultiplier tube. The intensity of these rays, which constitute the specimen beam, varies with the turbidity of the specimen.

"Rays that are reflected by the beam splitter fall on a front-surface mirror, from which they are reflected through a

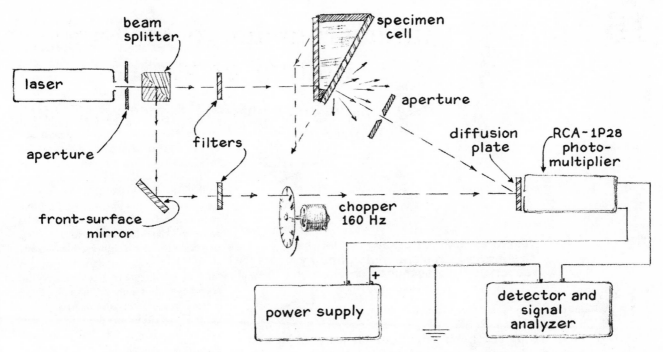

*Top view of elements of the turbidimeter designed by a group at the University of Rochester*

neutral-density filter and the apertures in a motor-driven chopper. The chopper interrupts the beam 160 times per second. Rays of the flickering beam recombine with those of the specimen beam at the photocathode.

"The output of the photomultiplier tube is a unidirectional electric current that varies in amplitude 160 times per second. A circuit separates the output into two parts: a direct-current signal and an alternating-current signal [*see bottom illustration on this page*]. The direct-current signal represents the

sum of the current generated by light transmitted through the specimen plus the direct-current component of the chopped reference beam. It is measured by meter No. 1. The amplitude of the alternating-current signal is proportional to the intensity of the reference beam. It is measured by meter No. 2.

"The electronic circuit consists of a network that includes three operational amplifiers, which boost the output of the photomultiplier to a usable level. The alternating-current and direct-current components are amplified independently

for display on separate meters. The filter that separates the alternating current from the direct current consists of an inductance of 1.4 henrys and a capacitor of .7 microfarad connected in series with the input of the reference-signal amplifier, $A_3$. The filter transmits maximum current at a frequency of 160 hertz. Both the capacitor and the inductor are available commercially.

"The frequency at which the chopper operates is not critical. Choppers can be improvised to operate at other frequencies. For example, a synchronous

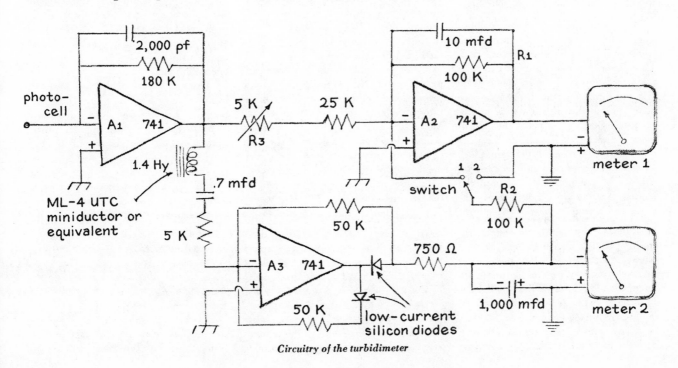

*Circuitry of the turbidimeter*

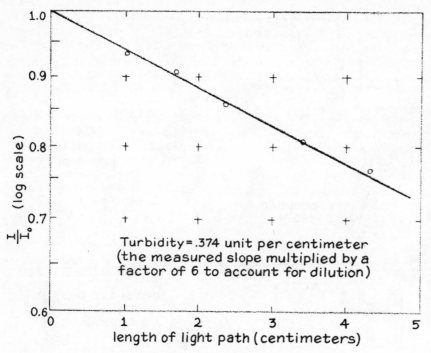

Turbidity=.374 unit per centimeter
(the measured slope multiplied by a
factor of 6 to account for dilution)

*Turbidity of a water sample from Lake Ontario*

motor that operates at 3,600 revolutions per minute could be fitted with a disk that contains two apertures to chop the beam at a frequency of 120 hertz. The input circuit of the reference-beam amplifier could be tuned to this frequency by retaining the 1.4-henry inductance and increasing the capacitor to approximately 1.25 microfarads. The values of capacitors and inductances that resonate at other chopping frequencies can be calculated by the formula $1/6.28 \times (L \times C)^{1/2}$, in which $L$ is the inductance in henrys and $C$ is the capacitance in farads.

"The experimenter is primarily interested in the intensity of the transmitted beam as it is displayed on meter No. 1. This meter, however, responds to both the direct current of the specimen beam and the direct current in the reference beam unless the instrument is appropriately adjusted. To make the adjustment connect switch No. 2 to the position shown in the circuit diagram, block off the specimen beam and vary the resistance of potentiometer $R_3$ until the pointer of meter No. 1 falls to zero. Unblock the specimen beam. The adjust-

ment need be made only once a day.

"The electronic components of the circuit are quite stable. The switch is not strictly necessary. It merely serves as a check for determining that the electronic components are working properly. When the switch is in position 1 and the circuit is properly adjusted, the alternating-current portion of the reference signal is subtracted at the input junction of the $A_2$ amplifier, because operational amplifiers reverse the polarity of signals. When the switch is turned to the other position, at which the output of the reference-signal amplifier is connected to ground, meter No. 1 displays the sum of the reference and specimen voltages instead of the difference.

"After the instrument has been adjusted meter No. 1 indicates voltage proportional to the intensity of the beam transmitted through the specimen cell. Meter No. 2 similarly indicates voltage proportional to the intensity of the light source. An easy way to use the instrument is to fill the cleaned specimen cell with clear water and record the ratio of the two meter readings: $M_1/M_2$ (water). The specimen is then transferred to the

cell and a second set of readings is made: $M_1/M_2$ (specimen). The transmission of the specimen, as calibrated against clear water, is then expressed by the ratio, $(M_1/M_2)$ (specimen)/$(M_1/M_2)$ (water). The apparatus automatically compensates for changes in the intensity of the light source, changes in the sensitivity or gain of the photomultiplier and most other variables that degrade the results of extinction turbidimeters.

"Ideally a specimen of water that is well mixed should be uniformly turbid, and a transmission measurement for one path length is sufficient to determine the turbidity. Experience suggests, however, that a series of turbidity measurements should be made through columns of fluid of varying length to improve the accuracy and as a check that only single scattering processes are involved. A plot of the logarithm of the transmission versus the length of fluid traversed should be a straight line. Systematic deviation from a straight line indicates multiple scattering and a difficulty in calculating the turbidity.

"Our specimen cell was designed expressly with this requirement in mind. As mentioned, it consists of a right triangular prism. The vessel is mounted to a traversing table with the faces of the prism perpendicular to the plane in which the transmitted and reference beams are located. The prism is moved in a direction parallel to the hypotenuse of the triangle. The light continues to enter the specimen cell at a right angle to the vertical face as the specimen cell is displaced in the plane of the hypotenuse, but the length of the path traversed by the light beam varies directly with the position of the cell. The cell is moved along the traversing table by a screw mechanism calibrated in increments of one millimeter.

"The four pieces of the cell were cut from ordinary window glass. Three rectangular sides were glued to a base piece to form a right-triangular prism. Before making the strips with a glass cutter of the wheel type we inspected the sheet and chose areas that were free of bubbles and striations. Surfaces of optical quality are not required because the measurements involve intensity, not

cathode                                                                                                anode

typically −750 v                    R = 50 K each                                                        $A_1$

*Circuitry of the photomultiplier tube*

image formation. Any convenient angle can be made between the hypotenuse and the side through which the beam enters. Our vessel was assembled with Sauereisen Adhesive Cement, an acid-proof cement that is distributed by the Fisher Scientific Company in Rochester, N.Y. Doubtless any epoxy cement would work as well if the cell is cleaned only with detergent.

"We clean the cell after most measurements with a solution of chromic acid made by dissolving from 30 to 50 grams of sodium dichromate or potassium dichromate in a liter of concentrated sulfuric acid. A comparable cleaning preparation known as Chromerge is available commercially. It must be similarly diluted with concentrated sulfuric acid.

"To clean the cell wet the inside surfaces of the glass with the acid mixture, let the cell stand for five minutes and then rinse it thoroughly with filtered water. We prepare the rinsing water with a Millipore filtering apparatus. In our opinion cells need not be cleaned with chromic acid unless extreme accuracy is desired. For the routine assessment of turbidity in natural streams the vessel can be cleaned with any good household detergent.

"We found that the results of our transmission measurements were reproducible to within 1 percent. We also found that we had to dilute specimens from rivers and lakes about six to one, even when they were taken from calm water, to avoid significant multiple scattering of the light and a corresponding loss of accuracy. Usually severe multiple scattering occurs when turbidity attenuates the light more than 15 percent.

"Our results were checked by theoretical calculations of the turbidity of known samples and also by experiment. A typical experimental check was made with a specimen prepared by suspending polystyrene spheres two microns in diameter in filtered distilled water. We counted the number of spheres in a known volume of solution by means of a microscope and estimated the concentration to be between 100,000 and 200,000 particles per cubic centimeter. Subsequent measurements with our turbidity meter indicated an actual concentration of approximately 130,000 spheres per cubic centimeter. The test is based on the fact that concentrations of spheres of uniform diameter attenuate light by a predictable amount.

"Spheres of this kind are available commercially from a number of suppliers, including the Dow Chemical Company (Midland, Mich. 48640) and Particle Information Service (600 South Springer Road, Los Altos, Calif. 94022). The spheres come in the form of a concentrated solution. Dilute the solution by adding it to distilled water. This procedure prevents the spheres from clumping. We used one drop of the sphere solution in 250 milliliters of water for a 'stock' solution that was diluted further to conduct scattering experiments. We positioned the specimen cell near the bottom limit of its traverse for the maximum optical path length and diluted the sphere solution until the attenuation was less than 15 percent.

"To verify the concentration with a microscope we counted the particles against a translucent background in the form of a grid. If the focal length of the objective lens is known, the depth of the field can be computed. If the concentration is low, the microscope views particles in a single plane. We counted the number of particles per unit length of the field in one direction and cubed the result to find the concentration per unit volume.

"A polydisperse specimen was obtained at a depth of eight meters in Lake Ontario near the inlet of the Genesee River. Turbidity was measured after the specimen had been diluted six to one with distilled water. The results were expressed as a graph by plotting the logarithm of the ratio of the transmitted light ($I$) divided by the unattenuated intensity of the source ($I_0$) against the length of the path through the specimen that was traversed by the light beam [see top illustration on opposite page]. Turbidity is equal to the slope of the graph. In this example it amounts to .374 unit of turbidity per centimeter of path length.

"The apparatus can be used to measure the distribution of sizes and concentrations of particles, provided that the particles are assumed to be spherical in form so that they settle at a predictable rate. The rate at which the intensity would change with time could then be related to particle diameter.

"The physical details of the construction are left to the resources of the experimenter. The optical parts of our instrument were mounted to a metal base by rugged fixtures that were made with machine tools. Most of the parts could be made of wood or plastic. We enclosed the entire apparatus in a light-proof housing so that measurements could be made in a normally lighted room.

"Our apparatus is not readily portable. Moreover, the laser and the light chopper require a source of 115-volt, 60-hertz power. For this reason the apparatus cannot be used in the field. It should be possible to modify the apparatus for use with battery power by exchanging a certain amount of accuracy and precision for portability. For example, a photodiode or phototransistor could be substituted for the photomultiplier. High-intensity lamps that operate on 12 volts, together with color filters and collimating lenses, could replace the laser. It should also be possible to substitute for the motor-driven chopper an oscillating chopper improvised from a 60-hertz resonant switch of the kind formerly used in the inverter circuits of automobile radios."

# INSTRUMENTS OF DISPERSION

# IV

# IV INSTRUMENTS OF DISPERSION

## INTRODUCTION

T he articles that follow present four basic instruments designed to disperse light into its component wavelengths (colors). A spectroscope uses a prism for this purpose; several other instruments employ diffraction gratings. If the instrument permits the separation to be viewed directly with the eye, it is called a spectroscope. If a photographic film is used, the instrument is called a spectrograph. If a photometer is substituted for the film, the instrument is a spectrophotometer. If the instrument is designed for observations of solar emissions, it becomes a spectroheliograph.

Why would one want to disperse light? When Newton held a prism in sunlight, his goal was to relate colors to white light. His simple experiment revealed that white light is a mixture of the basic colors in approximately equal amounts. Now, several centuries later, that simple separation is put to more practical uses. If a material is to be analyzed objectively for its color, it cannot merely be examined by a casual observer, for the human observer is far from objective in interpreting colors. (Indeed, some sections of psychology are devoted to understanding how a person perceives color.) As Stong describes in one of the following articles, a more objective interpretation can be made with a spectrophotometer. The light transmitted through or reflected by the object is dispersed, and a photometer is passed through the light to record its intensity according to wavelength.

As dispersion techniques became more refined in the nineteenth century, a puzzling feature was noticed in the light from certain elements. When sodium was burned and observed with the spectroscope, for example, an emission at a particular place in the yellow appeared to be far brighter than the general background. Further refinement in the spectroscopy revealed that the sodium emitted light at only certain discrete wavelengths. Why this was so was a complete mystery. According to the concept of matter predominant at the turn of the century, burning sodium should emit light uniformly across the entire visible spectrum.

The answer to the riddle of the emission lines came early in this century when Niels Bohr revealed the quantization of the hydrogen atom. Although atomic theory has become far more complex since then, the central feature of the theory remains that the emission and absorption of light by atoms must occur at certain wavelengths. The wavelengths corresponding to each element are so precise that the element can be identified by them.

This fact is applied in much of modern astrophysics. When light from the sun or from some other star is dispersed through a special spectroscope designed for astronomical work, an astronomer can identify which elements are in the atmosphere of the star. Nuclear reactions in the star's core provide light through a broad range of wavelengths. As this light passes through the stellar atmosphere, gases absorb it at the wavelengths characteristic of the elements present there. Hence, the starlight is darker at particular wavelengths, allowing an astronomer to identify the star's atmospheric composition by comparing the missing wavelengths with those known to be absorbed by the various elements.

Much of astronomical spectroscopy is like a Sherlock Holmes story. A handful of clues provide the only means for identifying steller characteristics. The science has become so well developed that the modern-day Holmes can make the identification even from the faint light of a distant star.

# 17 Ocular Spectroscope

*A spectroscope for a telescope that separates colors in starlight*

December 1952

As early as 1814 Joseph von Fraunhofer, the father of astrophysics, placed a prism before the 1.2-inch lens of a theodolite and mapped the dark lines of the solar spectrum he saw, designating them with the now familiar letters. These are the Fraunhofer lines that give the stars the separate individualities of different human faces—individualities that are but dimly realized by those who observe only with a telescope. Unlike the telescope, the spectroscope reaches into a star and takes a sample of it. Paul W. Merrill of the Mount Wilson and Palomar Observatories has said that studying a star by telescope is like "trying to guess the contents of a book from its size, weight and general appearance; while a spectroscopic observation is opening the book and reading it through line by line."

Today astrophysics, which deals with the physical and chemical characteristics of the stars, is the largest branch of astronomy; in fact, the astrophysical tail now wags the astronomical dog. Yet not one amateur astronomer in 100 attaches even a simple spectroscope to his telescope or seeks to become an amateur astrophysicist. True, much of astrophysics is abtruse, but not all of it. Getting started has been the chief obstacle.

A simple way to get a start in astrophysics is to build the little ocular spectroscope described by Roger Hayward's drawing on the next page. With it you can study the spectra of the brightest stars, including the sun, directly or as reflected by the moon. This spectroscope will show the more prominent lines of the solar spectrum when held in the hand and aimed at the sun. But when you insert it in the telescope in place of the eyepiece, take care not to look through it directly at the sun, for that can make you blind. Without the telescope the spectroscope may also be used on light sources such as neon tubes, a salted gas flame or a welder's iron arc.

It is called an ocular spectroscope because its diameter is uniform with the standard telescope ocular, or eyepiece. It is kept with the set of oculars and adds variety to their use. Its multicolored diffraction-grating spectra will also serve to satisfy the astronomically unsophisticated visitors whom all telescope owners occasionally have to entertain and who, seeing only with their eyes and not with their understanding, fail to be impressed. The ocular spectroscope will make your Aunt Emma say "Ah!" even though she may never have heard of Kirchhoff's three laws of spectrum analysis.

The midget spectroscope was designed and built by Ernst Keil, an amateur astronomer and professional instrument maker at the California Institute of Technology in Pasadena, Calif. As an avocation he has from time to time designed and built little ocular spectroscopes, including one for James Fassero, the author of *Photographic Giants of Palomar*, who uses it in his lecture demonstrations with the 100-inch telescope at Mount Wilson.

The achromatic lens of about two inches focal length may be obtained from war surplus for a dollar or two, or a plano-convex lens may be substituted with little optical loss. The only working dimension is the 1¼-inch outside diameter, a carefully machined sliding fit for the telescope drawtube. The other dimensions are those you choose. There are no "blueprints." Keil supplies only the little round gratings, which are replicas made by his own process, developed years ago and different from others. "The replica film," he writes, "is an integral part of the

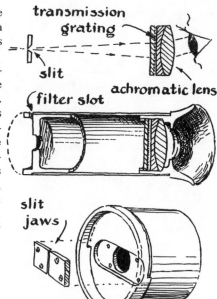

*A simple ocular spectroscope*

glass backing on which it is cast and is not a negative but a positive, giving the same distribution of light as the original." For a simple spectroscope a replica is as good as an original, and costs much less. The only way today to obtain a small original grating is to buy the costly laboratory spectrograph of which it is a part.

"The slit," Keil writes, "consists of two steel jaws made with care, their razor-sharp edges perfectly straight; see *Amateur Telescope Making*, page 248. The better the jaws, the sharper and more distinct will be the spectrum lines. Their separation will depend upon the brightness of the star observed, but .01 inch should be suitable.

"The light from a star is gathered by your telescope and focused in the plane of the slit jaws. Entering the slit, it passes through the transmission grating, which disperses it into its colors, then through a lens that collimates the light (making it parallel) and magnifies the spectrum. In this spectroscope the grating is put behind the collimator, instead of in front of it, to protect the grating. Actual trial will prove that in this simple spectroscope it makes no difference on which side of the grating the collimator is placed, for the spectroscope is not intended for serious scientific research but only for demonstrating the elementary principles of spectroscopy.

"To put the instrument in operation, first rotate the grating-lens unit, which must have a sliding fit inside the outer tube, until the grating lines are parallel with the slit. Then slide it in or out until the slit is in sharp focus. Insert it in the telescope and move it in or out until brilliant spectra appear.

"One available replica has 7,500 lines per inch and makes a spectrum of great intensity but comparatively small dispersion. Another, with 15,000 lines per inch, has about twice the dispersion of the first but a less brilliant spectrum.

"The slot on the front of the spectroscope is at right angles to the slit and of such a depth that a filter placed in it will cover one half of the slit. Two spectra, one above the other, will then be seen simultaneously—one the original, the other an absorption spectrum. Gelatin filters may be had from the Eastman Kodak Company or you can use red or blue cellophane, obtainable at photography stores."

A less serious addition to the amateur telescope owner's set of eyepieces was made by Alan R. Kirkham. He built a Tolles solid eyepiece lens of crystal quartz which is doubly refracting and produces two images. Thus he could always reveal a "secret area" of the sky where all the stars were double. An "eyepiece" built by Leo J. Scanlon consisted of a spinthariscope mounted inside an eyepiece shell. This is a particle of radium compound in front of a fluorescent screen of zinc sulfide, set behind the magnifying eye lens. Thus he could always show "exploding universes" through his telescope.

# 18 Bunsen Spectroscope

## Reconstructing the spectroscope that initiated modern spectroscopy

June 1955

Atoms and molecules, when struck a sharp blow by a hammer of atomic dimensions, ring like bells. The ear is not sensitive to the electromagnetic waves they emit, but the eye is. All light originates in this way. Just as every bell makes a characteristic sound of its own, depending upon its size and shape, so each of the hundred-odd kinds of atoms and their myriad molecular combinations radiate (or absorb) light of distinctive colors. The instrument physicists use to sort out the colors, and thus identify substances, is the spectroscope. This powerful instrument is relatively simple in principle and not difficult for amateurs to build. Yet it is one of the most useful tools in the scientist's kit. During the first half of this century it has helped to answer an incredible number of scientific questions—more than the telescope and microscope combined.

Accordingly, spectroscopy has become a specialized and important branch of physics. Any substance, whether a piece of cheese or a rusty hairpin, can be made to emit light by heating it. When the resulting light is sent through a spectroscope, the rays separate into lines or bands of colors which not only tag the responsible atoms but may reveal many secrets of how they are combined.

Light waves are only about one 50,000th of an inch long. With a good spectroscope, however, you can measure their size within a few trillionths of an inch, or about a billionth of the thickness of the paper on which this magazine is printed. With this information as a guide, chemists have learned how to take molecules apart and reassemble them into substances with new and desirable properties. About 20 kinds of molecules have been manufactured in

the laboratory for every one chemists have identified in nature. The tool contributing most to this analysis and synthesis is the spectroscope.

Isaac Newton laid the foundations of spectroscopy when he observed that a prism bends rays in the blue end of the spectrum more than those nearer the red end. On February 6, 1670, Newton wrote Henry Oldenburg, then secretary of the Royal Society:

"To perform my late promise to you, I shall without further ceremony acquaint you that in the year 1666, I procured me a triangular prism, to try therewith the celebrated phenomena of colors. And in order thereto, having darkened my chamber and made a small hole in my window-shuts, to let in a convenient quantity of the Sun's light, I placed my prism [so that the ray] might be refracted to the opposite wall. It was at first a very pleasant divertissement to view the vivid and intense colors produced thereby, but after awhile, applying myself to consider them more circumspectly, I became surprised to see them in oblong form which according to the laws of refraction, I expected should have been circular. . . . I took two boards and placed one of them behind the prism at the window so that the light might pass through a small hole made in it for the purpose and fall on the other board which I placed about 12 feet distance. Then I placed another prism behind this second board so that the light trajected through both boards might pass through that also. . . . This done, I took the first prism in my hand and turned it to and fro slowly about its axis so much as to make the several parts of the image cast on the second board successively pass through the hole in it that I might observe to what places on the wall the second prism would refract them. I saw . . . that the light on the (violet) end did in the second prism suffer a refraction

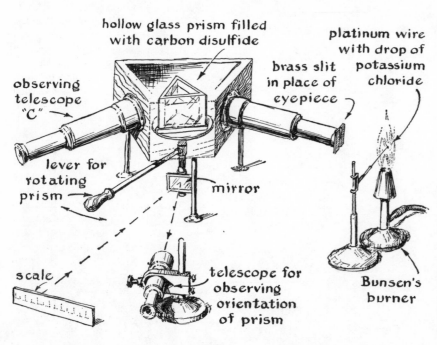

*A spectroscope reconstructed according to the directions of Robert Bunsen*

considerably greater than the light tending to the other end (red) . . . and that according to their particular degrees of refrangibility they were transmitted through the prism to divers parts of the opposite wall."

With this demonstration Newton's service to spectroscopy came to an end. He apparently failed to see any of the fine detail which gives the spectrum its significance. Nevertheless he subsequently stated that the sorting of the colors could be carried further by the use of improved prisms and lenses.

It is difficult to understand why Newton did not reduce his spectroscope to convenient laboratory form. He almost had it in principle. Yet he and his successors were content to work in darkened rooms for nearly two centuries. Even Joseph von Fraunhofer's epoch-making discovery of the dark absorption lines, which split the solar spectrum into thousands of parts, was made with a setup that filled his laboratory. To view the slit in the "window-shut," beyond his prism, Fraunhofer used a theodolite telescope placed behind the prism—an improvement which enabled him to make good measurements of the angles through which the light was bent. The clear view of the slit thus afforded disclosed "an almost countless number of strong and weak vertical lines," which close examination proved were "in the sunlight." Fraunhofer could not explain the lines, but he made an accurate chart of about 700 of them and designated eight of the most prominent ones by the letters A to H, by which they are still known.

The meaning of the dark lines remained a mystery until 1859, when they were explained by the Heidelberg physicists Robert Bunsen and Gustav Kirchhoff. They made the profound discovery that gases through which a ray of light passed would absorb certain narrow portions or colors of the light. The absorptions were signaled by dark lines in the spectrum. They also demonstrated that if the absorbing gas itself was heated to incandescence, then the dark lines of absorption became bright lines of emission, which stood out on the dark spectral band if there were no other source of light.

In the course of their experiments Bunsen and Kirchhoff reduced the size and design of the prism-type spectroscope to substantially its present form. Amateurs who would like to experiment with one may enjoy building the instrument according to the directions written by Bunsen 96 years ago.

"It is well known," he wrote, "that many substances have the property when they are brought into a flame, of producing in its spectrum certain bright lines. We can base a method of qualitative analysis on these lines that greatly broadens the field of chemical research and leads to the solution of problems previously beyond our grasp.

"The gas lamp previously described [Bunsen's gas burner] gives a flame of high temperature but low luminosity. Into this flame we introduced for investigation a small quantity of chlorate of potassium which had been recrystallized six or eight times. The apparatus we have used for investigating spectra is shown [*see illustration on the opposite page*]. The box [holding the prism] is blackened on the inside. Its two inclined sides carry two small telescopes. The ocular of the one facing the test flame is replaced by a plate in which is a slit formed by two brass blades. The burner is placed before the slit. The end of a fine platinum wire, bent into a small loop and supported by an apparatus stand, passes into the flame; on this hook is melted a globule of the chloride previously dried. Between the objectives of the two telescopes is placed a hollow prism with a refracting angle of 60 degrees and filled with carbon disulfide. The prism rests on a brass plate that can be rotated by a vertical shaft. The shaft carries on its lower end a mirror, above which an arm attaches which serves as a handle for turning the prism and mirror. Facing the mirror is another small telescope arranged to give an image of the horizontal scale, placed a short distance away. By rotating the prism, one can make the entire spectrum of the flame pass before the vertical cross-hair in the ocular of the viewing telescope. To every point in the spectrum, there corresponds a certain reading of the scale."

With this instrument Kirchhoff and Bunsen made the series of elegant investigations which founded the modern science of spectroscopy. Their explanation of the Fraunhofer lines and discovery of the elements cesium and rubidium inspired scientists all over the world to take up the new field of investigation and raised public interest in science to a high pitch.

Unfortunately Bunsen omitted one critical detail of construction that has plagued instrument makers ever since. He failed to specify the kind of cement he used to join the glass slabs of his hollow prism and seal in the foul-smelling, volatile, explosive and poisonous carbon disulfide.

R. B. Nevin of Christchurch, New Zealand, has made Bunsen-type prisms with wax as the sealer. He describes them as follows:

"My 60-degree prisms are made of eighth-inch plate glass which has never heard of such a thing as a figure. The glass is cut into pieces 2.5 by 2 inches, accurately oblong. Using a carborundum stone, I bevel the long edges slightly on one side at an angle of 30 degrees to the horizontal, and similarly bevel both sides of the shorter edges. The three slabs are then assembled on the bench as an equilateral triangle, the bottom one being stuck with anything handy if it won't stay put. A few bits of sealing wax are put in the top groove, and a gas flame is gently wafted along the glass until the wax melts. It is then spread with a match stick and given more heat until liquid. Next it is coaxed firmly but gently into the groove with the stick. More wax is applied until the groove is filled level with the glass edges. Then the whole is allowed to cool. A Bunsen burner flame about half on and just nonluminous is correct. If the flame is too intense, it heats the glass unevenly and cracks it. The procedure is repeated for the remaining two grooves. If the glass has been cut squarely and beveled properly, the edges will fit without adjustment. The wax holds to the ground edges tenaciously and is easy to flake off the polished surfaces where it is not wanted.

"The assembly is very strong for its dimensions. A triangular bottom for the box is then cut about a sixteenth of an inch larger than the assembled walls. The upper edges, upon which the walls will rest, are similarly ground, beveled and cemented to the previously completed subassembly. Gentle finishing touches with the flame to give a smoothly finished job can be applied to taste. The important precaution is: Do not rush the job. The sudden application of heat to one spot will crack the glass.

"When the prism is thoroughly cool, water can be poured into it, and when you look at a source of white light through it you will see all the colors of the rainbow. Glycerin in place of water will improve the prism's definition slightly, although its dispersion is about that of water and of crown glasses. It is significantly lower than flint glass. Perhaps the best liquid of all is carbon disulfide. Unfortunately it is a splendid solvent for sealing wax!

"That's about all there is to it. I whipped through a quite serviceable prism the other day in about 30 minutes of careful work—from cutting the glass to filling it with water."

Liquid prisms offer a number of distinct advantages to the amateur. They are easy and cheap to make, especially in large sizes. They also allow a wide choice of materials and dispersions. Their principal disadvantage is that the dispersive power of liquids varies greatly with changes in temperature. Moreover, temperature gradients within the liquid create inhomogeneities in dispersion with a consequent loss of resolution. The Fraunhofer lines blur and merge unless the liquid is maintained at uniform temperature.

Roger Hayward, after making the drawing of Bunsen's spectroscope shown here, volunteered a few practical tips out of his experience with liquid prisms. "Carbon disulfide," he wrote, "is terrible stuff. In addition to being smelly and explosive, it is a particularly insidious, chronic poison which can produce permanent damage to the spinal cord and brain. It is like playing with a bunch of uncaged cobras. Never handle it except under a ventilating hood.

"The best substitutes for carbon disulfide from an optical point of view are monobromonaphthalene, ethyl cinnamate (expensive), an aqueous solution of barium-mercuric bromide and oil of bitter almonds, in that order. The prism should be made with a glass bottom. The difference between the temperature coefficient of glass and that of brass makes a leakproof joint between the two difficult to achieve. Glass is the simplest material to use because the pieces can be ground to fit. Perhaps litharge and glycerin would make a good cement. Few liquids dissolve it.

"Those who wish to avoid the labor of building the two small telescopes used in this instrument may buy a pair of the popular-priced, low-power telescopes now on the market. I have a small spectroscope that uses a telescope and collimator with apertures of only three quarters of an inch. It will not separate Fraunhofer's D line of sodium but will resolve the yellow lines of mercury.

"Incidentally, Bunsen's way of rotating the prism to scan the spectrum makes you shudder. Of course he knew no better. Using fixed telescopes, he had to set the angles in a way which did not allow him to focus on any part of the spectrum at minimum deviation. Thus all measures of angles were made from an arbitrary, nonreproducible point. When provision is made for rotating the telescopes around the prism, the point of minimum deviation for a single line in the spectrum is easily found. Then rotation of one of the telescopes can be measured from such a position and a reproducible measure made.

"An easy way to make a good slit is with safety-razor blades. A thin, double-edged blade snapped in two in the middle gives a fine pair of jaws. It is a painstaking job to file a pair of slit jaws, even as short as a quarter inch. The slit should be only two or three thousandths of an inch wide if you want to see detail.

"As to light sources, a neon lamp makes a dandy. Some 40 clear lines are visible, from the yellow-green down to the deep red. I have a little lamp in a quartz envelope because there is another nice bunch of lines in the near ultraviolet. Incidentally, I built a quartz spectrograph back in 1938. All the parts were home-built—prism, mirrors, slit and all. It displays a photographable spectrum about nine inches long which has to be taken in two-inch bites. This instrument became the prototype of a commercial spectroscope which sells for about a kilobuck. Nothing has been published on it, but if readers express a hankering for a description, I should be pleased to write it up for this department."

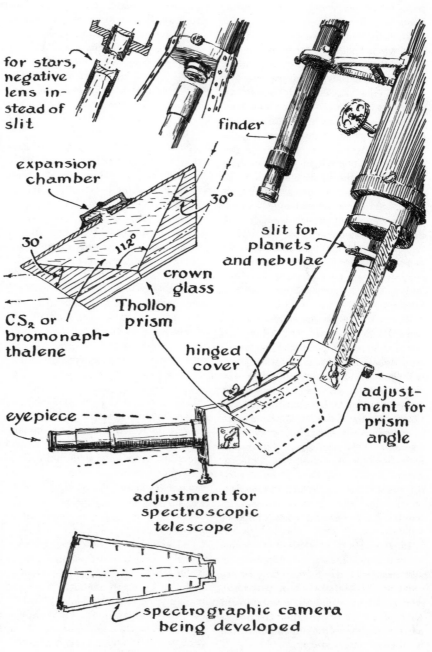

*A stellar spectroscope employing a liquid prism*

## Note on Making Liquid Prisms

*April 1956*

Last June this department carried a description of Robert Bunsen's original spectroscope, which employed a triangular glass box filled with carbon disulfide as the prism. Amateurs generally experience some difficulty in making liquid prisms. A Portuguese correspondent, Commander Eugenio C. Silva (whose observatory and 20-inch Cassegrainian telescope were described in this department in September, 1952) writes that he has found simple solutions for some of the problems mentioned in the Bunsen article.

"For some time," he writes, "I have been experimenting with liquid prisms and have uncovered some information which may be helpful to those who find themselves up against the characteristic difficulties of these devices. The main problem was finding an effective cement to hold the glasses together. It has to be some stuff which sets by cooling, does not require a volatile solvent and is quite insoluble in the liquid with which the prism is filled. For prisms filled with carbon disulfide or bromonaphthalene I have, after many trials, found a very good cement—shellac.

"In making the prisms I first heat the pieces of glass on an electric hot plate. The shellac is then applied to the edges to be joined and, as it begins to melt, is spread evenly with a needle. The hot plate must be regulated for the melting temperature of shellac and controlled so the shellac is neither burned nor comes to a boil. The work must then be cooled slowly to avoid cracking the glass. The joint is very strong, and after cooling the glass will break before the cement if you try to pull the elements apart.

"I have made two types of prisms. For medium dispersion I use the simpler ones of plate glass described by R. B. Nevin in "The Amateur Scientist" last June. For high dispersion I use the type developed by the French astronomer, Louis Thollon [*see drawing in the illustration on page 94*]. Though more difficult to make, the dispersion of Thollon prisms is four times greater than that of a conventional 60-degree prism of the same size. The single Thollon prism now in my telescope (filled with bromonaphthalene) easily resolves the 'D' doublet of sodium and a few faint lines can be seen between $D_1$ and $D_2$.

"The two crown-glass prisms of the Thollon design were cut from a glass slab about 30 millimeters thick and hand ground [*see lower drawings in the illustration below*]. The inside faces were fine ground and polished. Then the two prisms were cemented together with the plate glass top cover. The side faces (held together at the top) were then fine ground and the side plates cemented to the assembly. The rim around the base was fine ground and the base cover cemented in place. Finally outside faces were fine ground and polished in a plaster cradle. (These operations are much easier to perform than to describe.)

"The faces of the assembled prism were cleaned with a cotton swab dipped in alcohol, a good solvent for shellac.

"All liquid prisms must be fitted with a small expansion chamber to relieve the inner pressure when the fluid expands and contracts with changes in temperature. My prisms are equipped with the small metal chamber shown in cross-section. The chamber is left empty when the prism is filled. I did not provide for expansion in my first Thollon prism, which was filled with carbon disulfide. When I picked it up, the slight heat from my hand expanded the fluid and cracked the side plates! My prisms measure 30 millimeters in width and 40 millimeters in height. The glass covers measure two millimeters in thickness.

"As a stopper for the filler hole I use a small sheet of glass. The prisms cannot be heated on the hot plate after filling, so the glass cover is placed over the hole and cemented to the expansion chamber by means of a soldering iron.

"My prisms are now more than three years old and they have developed no leaks. It seems probable that they are as permanent as solid glass ones."

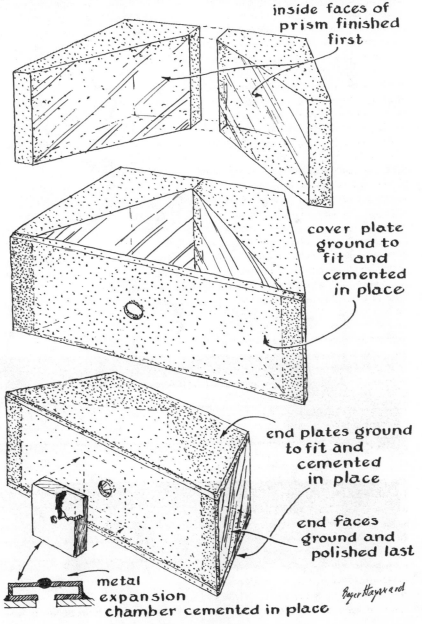

inside faces of prism finished first

cover plate ground to fit and cemented in place

end plates ground to fit and cemented in place

end faces ground and polished last

metal expansion chamber cemented in place

*Details of a liquid prism*

# 19

# Diffraction-Grating Spectrograph

*An inexpensive diffraction-grating spectrograph*

September 1956

Strip the astronomical telescope of its clock drive, film magazine, spectrograph and related accessories and you put it in a class with a blind man's cane. Like the cane, it informs you that something is out in front. Shorn of appendages, the telescope tells you next to nothing about the size, temperature, density, composition or other physical facts of the bodies which populate space. Not more than 20 celestial objects, other than comets, appear through the eyepiece as interesting patterns of light and shade. Only one, the moon, displays any richness of surface detail. All other bodies look much as they do to the naked eye. There is a greater profusion of stars, but as a spectacle the night sky remains substantially unchanged.

That is why the experience of building a telescope leaves some amateurs with the feeling of having been cheated. A few turns at the eyepiece apparently exhaust the novelty of the show, and they turn to other avocations.

Other amateurs, like Walter J. Semerau of Kenmore, N.Y., are not so easily discouraged. They pursue their hobby until they arrive at the boundless realm of astrophysics. Here they may observe the explosion of a star, the slow rotation of a galaxy, the flaming prominences of the sun and many other events in the drama of the heavens.

Semerau invested more than 700 hours of labor in the construction of his first telescope, described in this department in May, 1948. "I must confess," he writes, "that what I saw with it seemed poor compensation for the time and effort. That, however, overlooks other satisfactions: the solution of fascinating mechanical and optical problems. Considered in these terms, that first instrument was the buy of a lifetime."

Semerau soon decided, however, that he had to have a larger telescope equipped with devices to gather more information than his eye could detect.

Accordingly he went to work on a 12½-inch Newtonian reflector, complete with film magazine and four-inch astrographic camera. Both were assembled on a heavy mounting with an electric drive, calibrated setting-circles and slow-motion adjustments. He could now not only probe more deeply into space but also do such things as determine the distance of a nearby star by measuring its change in position as the earth moves around the sun. To put it another way, he had made his "cane" longer and increased his control of it. When the sensitivity of modern photographic emulsions are taken into account, Semerau's new instruments were almost on a par with those in the world's best observatories 50 years ago.

During these 50 years, as Cecilia Payne-Gaposchkin of the Harvard College Observatory has pointed out, we have gained most of our knowledge of the physics of the universe. Most of this knowledge has come through the devel-

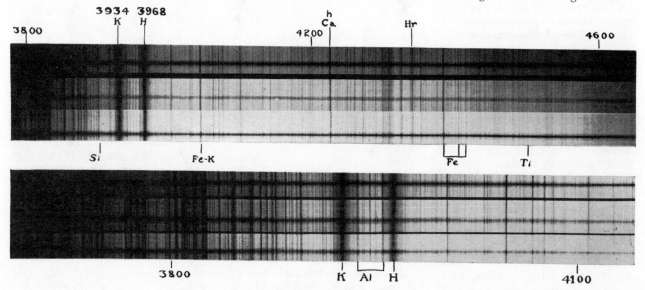

*Sunspot spectrograms made by Walter L. Semerau. The first-order spectrum is at the top; the second-order, at the bottom*

opment of ingenious accessories for the telescope which sort out the complex waves radiated by celestial objects.

Semerau now decided that he had to tackle the construction of some of these accessories and to try his hand at the more sophisticated techniques of observing that went with them. He went to work on a monochromator, a device which artificially eclipses the sun and enables the observer to study the solar atmosphere. Semerau's description of the apparatus, together with color photographs of solar prominences made with it, appeared in this department just a year ago.

Having built the monochromator, Semerau felt he was ready to attempt one of the most demanding jobs in optics: the making of a spectrograph. Directly or indirectly the spectrograph can function as a yardstick, speedometer, tachometer, balance, thermometer and chemical laboratory all in one. In addition, it enables the observer to study all kinds of magnetic and electrical effects.

In principle the instrument is relatively simple. Light falls on an optical element which separates its constituent wavelengths or colors in a fan-shaped array; the longest waves occupying one edge of the fan and the shortest the other. The element responsible for the separation may be either a prism or a diffraction grating: a surface ruled with many straight and evenly spaced lines. The spectrograph is improved by equipping it with a system of lenses (or a concave mirror) to concentrate the light, and with an aperture in the form of a thin slit. When the dispersed rays of white light are brought to focus on a screen, such as a piece of white cardboard, the slit appears as a series of multiple images so closely spaced that a continuous ribbon of color is formed which runs the gamut of the rainbow.

As previously discussed in this department [June, 1955], each atom and molecule, when sufficiently energized, emits a series of light waves of characteristic length. These appear as bright lines in the spectrum and enable the investigator to identify the chemical elements of the incandescent source. Similarly, the atoms of a gas at lower temperature than the source absorb energy at these characteristic wavelengths from light transmitted through the gas. The absorption pattern appears as dark lines. As the temperature of the source increases, waves of shorter and shorter length join the emission, and the spectrum becomes more intense toward the blue end. Thus the spectral pattern can serve as an index of temperature.

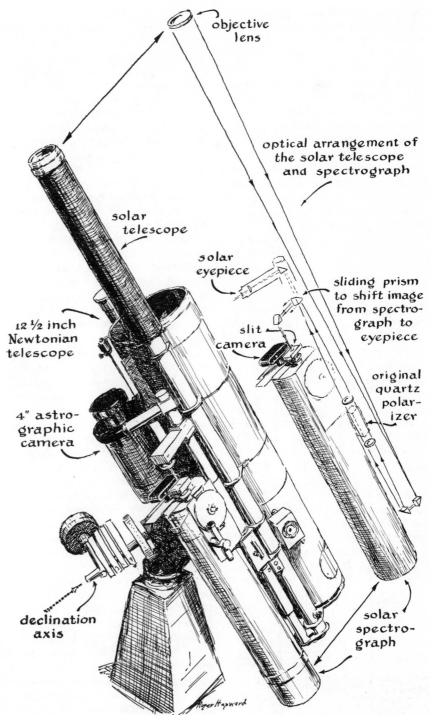

*Semerau's telescope, astrographic camera, monochromator and spectrograph*

The characteristic lines of a substance need not always appear at the same position in the spectrum. When a source of light is moving toward the observer, for example, its waves are shortened—the Doppler effect so frequently mentioned in this issue. In consequence the spectral lines of atoms moving toward the observer are shifted toward the blue end of the spectrum. The lines of atoms moving away are shifted toward the red. Velocity can thus be measured by observing the spectral shift.

When an atom is ionized, i.e., electrically charged, it can be influenced by a magnetic field. Its spectral lines may then be split: the phenomenon known as the Zeeman effect. Intense electrical fields similarly leave their mark on the spectrum.

These and other variations in normal spectra provide the astrophysicist with most of his clues to the nature of stars, nebulae, galaxies and the large-scale features of the universe. The amateur can hardly hope to compete with these observations, particularly those of faint objects. However, with well-built equipment he can come to grips with a rich variety of effects in the nearer and brighter ones.

"If you are willing to settle for the sun," writes Semerau, "you shuck off a lot of labor. A three-inch objective lens, or a mirror of similar size, will give you all the light you need. The rest is easy. Many amateurs have stayed away from spectroscopes because most conventional designs call for lathes and other facilities beyond reach of the basement workshop, and many are too heavy or unwieldy for backyard use.

"About four years ago I chanced on a design that seemed to fill the bill. My

*8" autocollimating mirror 45⅜" focal length*

*Optical train of the spectrograph*

employer, the Linde Air Products Company, a division of the Union Carbide and Carbon Corporation, needed a special spectroscope for industrial research and could not find a commercial instrument that met their specifications. The Bausch & Lomb Optical Company finally located a design that looked promising. As things worked out, it was adopted and is now on the market. My instrument, shown mounted on page 260, is a copy of that design.

"The concept was proposed by H. Ebert just before the turn of the century. The instrument is of the high dispersion, stigmatic type and employs a plane diffraction grating. As conceived by Ebert, the design was at least 50 years ahead of its time. In his day plane gratings were ruled on speculum metal, an alloy of 68 per cent copper and 32 per cent tin which is subject to tarnishing. This fact alone made the idea impractical. Ebert also specified a spherical mirror for collimating and imaging the light. Prior to 1900 mirrors were also made of speculum metal. It was possible but not practical to repolish the mirror but neither possible nor practical to refinish the finely ruled grating. Consequently a brilliant idea lay fallow, waiting for someone to develop a method of depositing a thin film of metal onto glass that would reflect light effectively and resist tarnishing. Then John Strong, now director of the Laboratory of Astrophysics and Physical Meteorology at the Johns Hopkins University, perfected a method of depositing a thin film of aluminum on glass.

"The process opened the way for many new developments in the field of optics. One of these is the production of high-precision reflectance gratings ruled on aluminized glass. Prior to being coated the glass is ground and polished to a plane that does not depart from flatness by more than a 100,000th of an inch. The metallic film is then ruled with a series of straight, parallel saw-tooth grooves—as many as 30,000 per inch. The spacing between the rulings is uniform to within a few millionths of an inch; the angle of the saw-tooth walls, the so-called 'blaze angle,' is held similarly constant. The ruling operation is without question one of the most exacting mechanical processes known, and accounts for the high cost and limited production of gratings.

"In consequence few spectrographs were designed around gratings until recently. About five years ago, however, Bausch and Lomb introduced the 'certified precision grating.' These are casts taken from an original grating. It is mis-

leading to describe them as replicas, because the term suggests the numerous unsatisfactory reproductions which have appeared in the past. The Bausch and Lomb casts perform astonishingly well at moderate temperatures and will not tarnish in a normal laboratory atmosphere. The grooves are as straight and evenly spaced as those of the original. The blaze angle can be readily controlled to concentrate the spectral energy into any desired region of the spectrum, making the gratings nearly as efficient for spectroscopic work as the glass prisms more commonly used in commercial instruments. Certified precision gratings sell at about a tenth the price of originals; they cost from $100 to $1,800, depending upon the size of the ruled area and the density of the rulings. Replicas of lesser quality, but entirely adequate for amateur use, can be purchased from laboratory supply houses for approximately $5 to $25.

"The remaining parts of the Ebert spectrograph—mirror, cell, tube, slit and film holder—should cost no more than an eight-inch Newtonian reflector. Depending on where you buy the materials, the entire rig should come to less than $100. By begging materials from all my friends, and keeping an eye on the Linde scrap pile, mine cost far less.

"There is nothing sacred about the design of the main tube and related mechanical parts. You can make the tube of plywood or go in for fancy aluminum castings, depending upon your pleasure and your fiscal policy. If the instrument is to be mounted alongside the telescope, however, weight becomes an important factor. The prime requirement is sufficient rigidity and strength to hold the optical elements in precise alignment. If the spectrograph is to be used for laboratory work such as the analysis of minerals, sheet steel may be used to good ad-

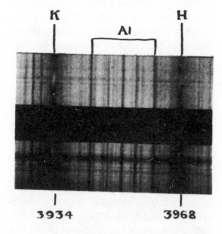

*Sunspot in the fourth spectral order*

vantage. For astronomical work you are faced with the problem of balancing rigidity and lightness. Duralumin is a good compromise in many respects. Iron has long been a favored material for the structural parts of laboratory spectrographs because its coefficient of expansion closely approaches that of glass. When mirrors are made of Pyrex, an especially tough cast iron known as meehanite has been used to counteract the effects of temperature variation.

"The optical elements of my instrument are supported by a tube with a length of 45 inches and an inside diameter of 8¼ inches [*see drawing on page 98*]. The walls of the tube are a sixteenth of an inch thick. The eight-inch spherical mirror has a focal length of 45⅝ inches. The grating is two inches square; it is ruled with 15,000 lines per inch. The long face of the saw-tooth groove is slanted about 20 degrees to the plane of the grating. The width of each groove is 5,000 Angstrom units, or about 20 millionths of an inch. Such a grating will strongly reflect waves with a length of 10,000 A., which are in the infrared region. The grating is said to be 'blazed' for 10,000 A. A grating of this blazing will also reflect waves of 5,000 A., though less strongly. These waves give rise to 'second-order' spectra which lie in the center of the visible region: the green. In addition, some third-order spectra occur; their wavelength is about 3,300 A. Waves of this length lie in the ultraviolet region.

"The angle at which light is reflected from the grating depends upon the length of its waves. The long waves are bent more than the short ones; hence the long and short waves are dispersed. A grating blazed for 10,000 A. will disperse a 14.5-A. segment of the first-order spectrum over a millimeter. My instrument thus spreads a 2,200-A. band of the spectrum on a six-inch strip of film.

"The film holder of my spectroscope is designed for rolls of 35-millimeter film. Light is admitted to the holder through a rectangular port six inches long and four tenths of an inch wide. By moving the holder across the port, it is possible to make three narrow exposures on one strip. This is a convenience in arriving at the proper exposure. The exposure time is estimated on the basis of past experience for one portion of the film; the interval is then bracketed by doubling the exposure for the second portion and halving it for the third.

"The most difficult part of the spectrograph to make is the yoke which supports the grating. Much depends on how well this part functions. It must permit

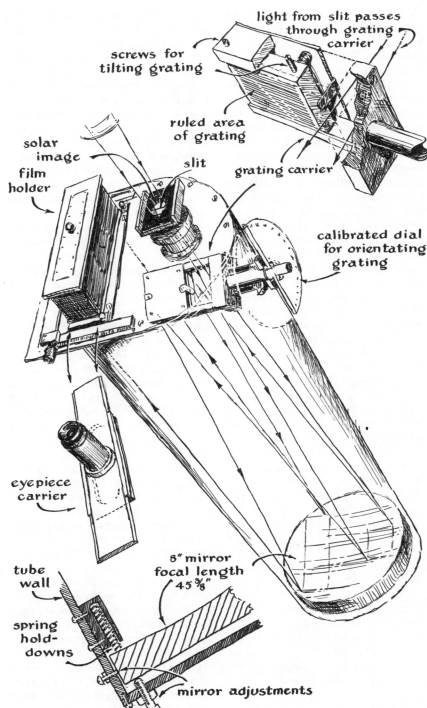

*Details of the spectrograph assembly*

the grating to be rotated through 45 degrees to each side, and provide adjustments for aligning the grating with respect to the mirror. The ruled surface must be located precisely on the center line of the yoke axis, preferably with provision for tilting within the yoke so that the rulings can be made to parallel the axis. In my arrangement this adjustment is provided by two screws which act against opposing springs, as shown in the drawing above. The pressure necessary to keep the grating in the

parallel position is provided by four springs located behind it. Two leaf springs, one above the other, hold the grating in place. The assembly is supported by an end plate from which a shaft extends. The shaft turns in a pair of tapered roller-bearings which, together with their housing, were formerly part of an automobile water-pump. A flange at the outer end of the housing serves as the fixture for attaching the yoke assembly to the main tube. It is fastened in place by two sets of three screws each, the members of each set spaced over 120 degrees around the flange. One set passes through oversized holes in the flange and engages threads in the tube. These act as pull-downs. The other set engages threads in the flange and presses against the tube, providing push-up. Adjusting the two sets makes it possible to align the yoke axis with respect to the tube.

"The shaft of the yoke is driven by a single thread, 36-tooth worm gear that carries a dial graduated in one-degree steps. The worm engaging the gear also bears a dial, graduated in 100 parts, each representing a tenth of a degree. The arrangement is satisfactory for positioning spectra on the ground glass or film but is inadequate for determining wavelengths.

"All plane gratings should be illuminated with parallel rays. Hence the entrance slit and photographic plate must both lie in the focal plane of the mirror. Small departures from this ideal may be compensated by moving the mirror slightly up or down the tube.

"The spectral lines of the Ebert spectrograph are vertical only near the zero order and tilt increasingly as the grating is rotated to bring the higher orders under observation. The tilting may be compensated by rotating the entrance slit in the opposite direction while viewing the the lines on a ground glass or through the eyepiece. The effect is aggravated in instruments of short focal length.

"The cell supporting the mirror, and its essential adjustments, are identical with those of conventional reflecting telescopes. If no cell is provided and the adjustment screws bear directly on the mirror—which invites a chipped back— then no more than three screws, spaced 120 degrees apart, should be used. This is particularly important if the screws are opposed by compression springs; more than three will almost certainly result in a twisted mirror.

"The film magazine is equipped with a 48-pitch rack and pinion, purposely adjusted to a tight mesh so each tooth can be felt as it comes into engagement. It is this arrangement that makes it possible to move the film along the exposure port and make three exposures on each strip of film. Lateral spacing during the racking operation is determined by counting the meshes. Although the magazine accommodates standard casettes for 35-mm. film, it is not equipped with a device for counting exposures. I merely count the number of turns of the film spool and record them in a notebook.

"The back of the magazine is provided with a removable cover so that a ground glass may be inserted as desired. It also takes a 35-mm. camera, a convenience when interest is confined to a narrow region of the spectrum such as the H and K lines of calcium or the alpha line of hydrogen. The back may be changed over to an eyepiece fixture which may be slid along the full six inches of spectrum. This arrangement provides for a visual check prior to making an exposure; it is especially helpful to the beginner.

"Care must be taken in illuminating the slit. If the spectrograph has a focal ratio of $f/20$ (the focal length of mirror divided by the effective diameter of grating), the cone of incoming rays should also approximate $f/20$ and the axis of the cone should parallel the axis of the mirror. The slit acts much like the aperture of a pinhole camera. Consequently, if the rays of the illuminating cone converge at a greater angle than the focal ratio of the system, say $f/10$, they will fill an area in the plane of the grating considerably larger than the area of the rulings. Light thus scattered will result in fogged film and reduced contrast. Misalignment of the incoming rays will have the same effect, though perhaps it is less pronounced. Baffles or diaphragms spaced every three or four inches through the full length of the tube will greatly reduce the effects of stray light, such as that which enters the slit at a skew angle and bounces off the back of the grating onto the film. The diaphragms must be carefully designed, however, or they may vignette the film.

"The components are assembled as shown in the drawing on page 99. The initial adjustments and alignment of the optical elements can be made on a workbench. An electric arc using carbons enriched with iron, or a strong spark discharge between iron electrodes, makes a convenient source of light for testing. The emission spectra of iron have been determined with great precision, and the wavelengths of hundreds of lines extending far into the ultraviolet and infrared (from 294 to 26,000 A.) are tabulated in standard reference texts. Beginners may prefer a mercury arc or glow lamp because these sources demand less attention during operation and emit fewer spectral lines which are, in consequence, easier to identify. The tabulations, whether of iron or mercury, are useful for assessing the initial performance of the instrument and invaluable for calibrating comparison spectra during its subsequent use.

"Recently I have been concentrating on the spectroscopic study of sunspots. To make a spectrogram of a sunspot you align the telescope so that the image of the sun falls on the entrance slit. The objective lens of my telescope yields an image considerably larger than the slit. The image is maneuvered, by means of the telescope's slow-motion controls, until a selected sunspot is centered on the slit, a trick easily mastered with a little practice. The spectrum is then examined by means of either the eyepiece or the ground glass. The spot is seen as a narrow streak which extends from one end of the spectrum to the other. The adjustments, including the width of the entrance slit, are then touched up so the lines appear with maximum sharpness.

"Successive spectral orders are brought into view by rotating the grating through higher angles. The upper spectrum on page 259 shows the first order. The one beneath is made in the second order. Note that although fewer lines per inch appear in the second order, there is no gain in resolution. Shifting the grating for the detection of a higher order is analogous to substituting eyepieces of higher power in a telescope. You get a bigger but proportionately fuzzier picture. The film magazine is substituted for the eyepiece and three exposures made in both the first and the second order. In many cases the range of intensity between the faintest and brightest lines exceeds the capacity of the film to register contrast. Three exposures, one estimated for the mid-range intensity and the other two timed respectively at half and twice this value, will usually span the full range.

"Gases in the vicinity of a sunspot often appear to be in a state of violent turbulence. At any instant some atoms are rushing toward the observer and others away. The spectral lines show proportionate displacement from their normal positions in the spectrum—the Doppler effect—and register as a bulge in the central part of the line occupied by the sunspot. This explains the dark streak extending through the center of the spectra reproduced on page 96.

"A portion of this same spectrum, photographed in the fourth order and enlarged photographically, appears on page 98. It includes the H and K lines of calcium, at wavelengths of 2,933 and 3,960 A. respectively. Observe that a segment in the center of each of these two lines—the segment representing the sunspot—is split. The light streak occupying the area within the split section is referred to as 'emission over absorption' and, in this instance, indicates the presence of incandescent calcium at an altitude of about 100,000 miles above a region of cooler matter in the spot. Had the glowing calcium been lower, its emission would have been absorbed by the intervening solar atmosphere and it would have photographed as a dark absorption line. My interpretation of this spectrogram is that a solar prominence, carrying incandescent calcium from the sun's interior, arched up and over the sunspot. We are looking down on top of it. Reconstructing such events from evidence buried in the myriad lines of spectra is an endless challenge and one of the hobby's many fascinations."

# 20

# Diffraction-Grating Spectrograph to Observe Auroras

*Auroral spectra made as part of the International Geophysical Year*

January 1961

Studies undertaken during the International Geophysical Year appear to have opened a new era in amateur astronomy. Before 1957 followers of this classic avocation occupied themselves largely with such time-honored activities as looking for new comets and observing changes in the brightness of variable stars. Today amateurs also track artificial satellites, patrol solar flares, electronically time the occultation of stars by the moon and participate in other investigations that were unknown a decade ago. The number of amateur astronomers has grown in proportion. Part of this burgeoning popularity can doubtless be explained by the novelty of the IGY projects. After all, artificial satellites are objects of uncommon fascination. But part can be ascribed to a new ease of acquiring instruments. Ten years ago most beginners could afford to own a telescope of reasonable power only by mastering the exacting art of shaping optical glass. Many of the current activities can be undertaken without any telescope at all. The required instruments can often be assembled with parts recovered from surplus apparatus or with those that fail to meet rigid specifications. For example, Walter A. Feibelman, a physicist of Pittsburgh, Pa., has used such parts to build a spectrograph for analyzing the light omitted by the aurora borealis. His design calls for little more than a pair of razor blades, an inexpensive replica diffraction grating, a conventional camera and a small achromatic lens. With a similar lens he has also made a telescopic camera to photograph the "green flash" and other curious optical effects that are associated with the setting sun.

"Like many amateurs," writes Feibelman, "I wanted to have a part in the IGY and volunteered to participate in the observation of auroras [see "The Amateur Scientist"; January, 1957]. The prospects for success did not seem too bright because one would not expect to see many auroras at the latitude of Pittsburgh. Moreover, industrial areas do not provide the most favorable seeing conditions. The project nonetheless turned

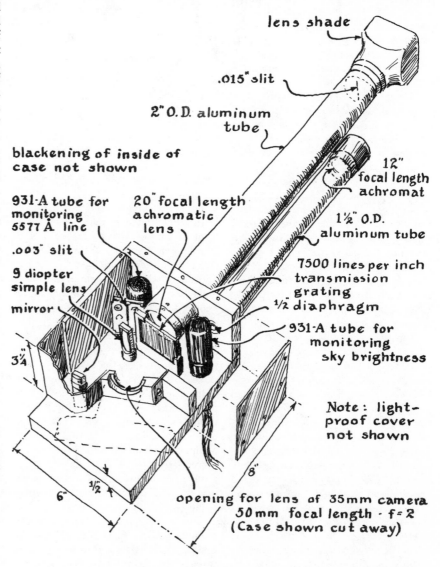

*An auroral spectrograph designed for amateur construction*

out to be highly rewarding. Beginning with the very large and bright display on the official opening night of the IGY, I observed and photographed a total of 45 auroras by November, 1960. Others were doubtless missed because of clouds, smog or bright moonlight. No two displays were ever exactly alike, and the changing patterns and colors were always fascinating to watch.

"In all I made over 700 black-and-white photographs of auroras, plus 20 or so in color. Most were taken with an old camera using 120-millimeter film. The exposures were 15 seconds or less at $f/2.9$. I used Royal-X Pan film developed in DK 60A. The pictures were so easy to make that the novelty soon wore off, and I began to cast about for something additional to do. Eventually I hit on the idea of building an aurora spectrograph. This instrument helps to distinguish among the several kinds of aurora and reveals the presence of displays that are invisible to the eye. It is a simple apparatus, but the literature disclosed scant information about its construction. With a few hints from astronomers I wound up designing my own.

"Aurora spectrographs include four basic parts: (1) a mechanical slit for admitting a thin ribbon of light to the instrument, (2) a lens for bending the diverging ribbon into a parallel beam, (3) either a diffracting or a refracting element for dispersing the collimated beam into its constituent colors and (4) a camera for recording the spectral pattern on photographic film. No telescope or other light-gathering device is used. None is needed to make auroral spectrograms. (No illumination would be gained by bringing a small patch of the glowing area to focus on the slit.) If desired, a telescope can be added. The instrument can then be used for photographing stellar spectra by focusing the image of a selected star on the slit.

"The structural arrangement is as follows. All of the optical parts are supported by aluminum tubing and an associated housing made of aluminum plates and sheet stock as depicted in the accompanying drawing [*preceding page*]. When in use, the instrument is mounted on a conventional camera tripod. A second camera, for making wide-field pictures, is mounted on an accessory bracket. The slit assembly consists of a cylindrical base that can be made of any convenient material such as aluminum, plastic or even hardwood. Two safety-razor blades are fastened to the base with screws as shown in the second drawing [*above*]. The cutting edges of the blades must be parallel and separated .015 inch. A rectangular hole, somewhat larger than the slit, is made in the base for admitting light to the lens. The slit assembly is mounted in one end of an aluminum tube. An achromatic lens 1¾ inches in diameter fits into the other end of the tube at the distance of its focal length from the slit—20 inches in the case of my instrument. Neither the diameter of the lens nor its focal length are critical. The diffraction grating is placed immediately behind the lens and at an approximate right angle to the beam, as depicted in the third drawing [*page 104*]. The grating and the back face of the lens are enclosed by a housing. A hole in the rear of the housing, which makes a light-tight fit with the lens barrel of the camera, admits the dispersed rays to the camera lens. The focusing adjustment of the camera is set at infinity. I use a 35-millimeter Retina camera equipped with an $f/2$ lens. The time required for exposures depends in part on the light-gathering power of the camera lens. In general, focal ratios substantially larger than $f/2$ are not satisfactory.

"For the dispersing element I prefer a transmission (*i.e.*, transparent) replica grating to a glass prism. A grating spreads the colors more than a prism does, and the dispersion is uniform throughout the spectrum. A spectrum made by a prism becomes increasingly crowded toward its red end. Spectral colors appear as parallel bands or lines that cross the ribbon-shaped pattern like the rungs of a ladder. Crowding obviously increases the difficulty of identifying interesting lines.

"Gratings are not without disadvantages, however. Whereas prisms disperse light into a single spectral pattern, gratings produce multiple spectra that overlap at the ends, somewhat like the 'ghost' images that appear on television screens. To minimize this effect, gratings are commonly 'blazed,' that is, ruled with V-shaped grooves that are tilted at a slight angle to deflect most of the light into one preferred image, or spectral 'order.' My grating is blazed to favor the first spectral order. The lines of this order accordingly show up more prominently in my photographs than those of higher orders. The lines of the higher orders are also more widely spaced than those of the first order, a further aid in distinguishing the first-order lines [see "The Amateur Scientist;" June, 1955, and April, 1958]. My grating was manufactured by the Bausch & Lomb Optical Co. and can be purchased through a dealer in scientific supplies.

"With this combination of grating and

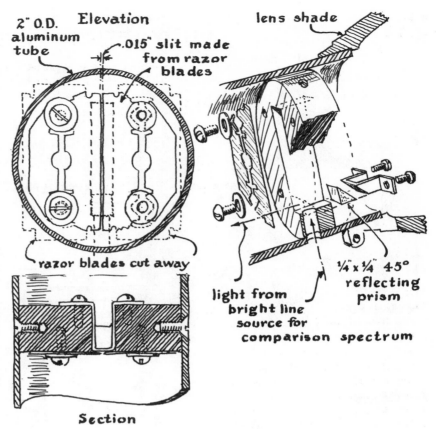

*Details of the slit for the auroral spectrograph*

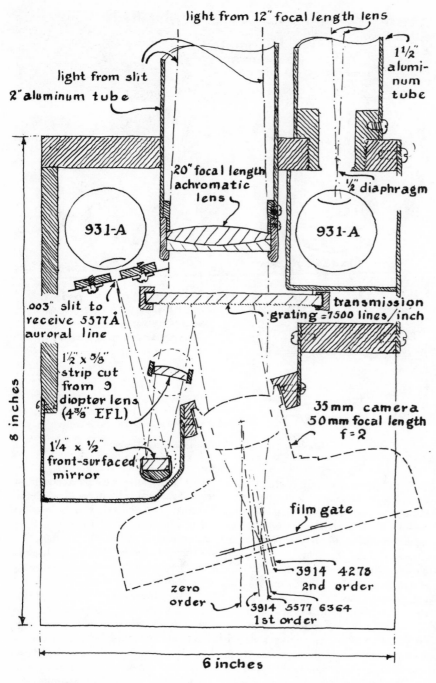

light from 12" focal length lens

light from slit

2" aluminum tube

1½" aluminum tube

20" focal length achromatic lens

½" diaphragm

931-A

931-A

.003" slit to receive 5577Å auroral line

1½ x ⅝" strip cut from 9 diopter lens (4⅜" EFL)

1¼ x ½" front-surfaced mirror

transmission grating ≈7500 lines/inch

35mm camera 50mm focal length f=2

film gate

zero order

3914 4278 2nd order

3914 5577 6364 1st order

8 inches

6 inches

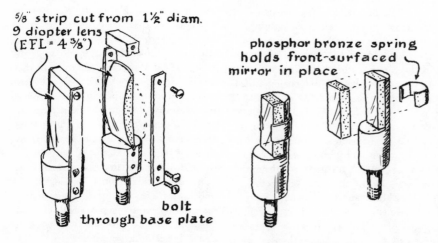

⅝" strip cut from 1½" diam. 9 diopter lens (EFL = 4⅜")

phosphor bronze spring holds front-surfaced mirror in place

bolt through base plate

*Plan view of the spectrograph* (top) *and details of the front-surfaced mirror* (bottom)

camera the exposure time required to photograph the brightest aurora averages about 10 minutes. Good photographs have been made with Ansco Super Hypan film. The best results are obtained, however, with spectroscopic emulsions such as Kodak 103a-F, which is specially sensitized for the far-red region of the spectrum. Spectrograms have been made of several auroras that were not visible, in some cases by extending exposures to four hours. The outcome of such attempts is a matter of pure chance. You have no way of knowing if an aurora is up there. So you just point the instrument toward the northern sky at night, open the shutter and go away. With luck, spectral lines characteristic of the aurora will appear when you develop the film.

"For many years auroral spectra presented investigators with something of a mystery. Certain of the lines could not be reproduced by light sources then available in the laboratory, and came to be known as 'forbidden' lines. The brightest of the forbidden lines appears in the green portion of the spectrum; it is formed by light waves that measure 5,577 angstrom units in length. This line is emitted when highly ionized atoms of oxygen in the upper atmosphere release a finite part of the energy acquired by collision with particles hurled into space by the sun. The 5,577-angstrom line is observed to some extent in all auroras. When intense, it accounts for the characteristic green color of most displays; when weak, the auroras look pale white or gray. Also characteristic are two lines at 6,300 and 6,364 angstroms in the red part of the spectrum; these also are emitted by ionized oxygen. Still another relatively prominent line is found at 3,914 angstroms in the extreme violet. It is emitted by ionized nitrogen. Weaker lines also appear throughout auroral spectra. Most are associated with the emission of energy from oxygen and nitrogen, although occasional spectrograms show a well-defined line at 6,563 angstroms that has been identified with hydrogen.

"The relative intensity of the spectral pattern varies with the intensity, height and type of the display; the temperature of the upper air; the state of the earth's magnetic field, and related environmental factors. Quite often spectral lines also show variation in brightness along their length. Auroras in the early evening or morning, when part of the display is in the earth's shadow and part in the sunlit upper atmosphere, are of particular interest because not all of the accompanying effects are understood in detail. They are investigated by pointing the spectrograph toward the area in question. The techniques of analyzing spectrograms to determine the height of a display or the temperature of the air in a selected region are described in the literature [*see page 144 of 'Bibliographies'*].

"Several optional features were built into the basic instrument to facilitate its operation. A small right-angle prism was added at the lower edge of the slit so that a comparison spectrum can be photographed simultaneously as an aid in the identification of unknown lines. Light from a small neon bulb is directed against the lower face of the prism for about 10 seconds. Any miniature neon bulb can be used. Neon emits particularly strong lines at 5,400, 5,852 and 6,402 angstroms. This accessory is not too important, because the 5,577 lines can always be recognized, as is apparent in the accompanying spectrogram [*top of this page*]. Incidentally, the lines may be sharpened by making the slit narrower than .015 inch. Sharpness is gained at the cost of longer exposure, however, because the light admitted to the film varies inversely with the width of the slit. Moreover, a narrow slit emphasizes the higher-order images.

"To aid in estimating exposures and plotting variations of auroral brightness a 931-A photomultiplier tube was incorporated in the instrument. The photocathode is illuminated by a small telescope equipped with an objective lens of 1½ inches diameter and 12 inches focal length. The incoming light is focused on the photocathode through a diaphragm 1/2 inch in diameter that limits the illumination to a patch of aurora about five degrees in diameter. The output of the phototube actuates a microammeter. Brightness is indicated simply by the movement of the pointer. If an ultrasensitive meter is used, so that a current of 10 microamperes drives the pointer to full scale, no amplifier is necessary. If not, the circuit must include a direct-current amplifier. Warning: The 900-volt supply for the photomultiplier can be lethal if handled carelessly. The tube and all parts of the circuit must be well insulated from the metal housing; no bare wires or terminals can be ex-

posed. The instrument is used in darkness, a condition that invites accidents.

"A meter of the pen-recording type would be preferable because it would plot a continuous graph of brightness against time automatically. I have not, however, invested in one. The accompanying graph [*bottom of this page*] of a typical aurora was plotted by hand from periodic meter readings. The display was of medium intensity and showed a strong peak at midnight. The spectrograph was pointed due north and about 20 degrees above the horizon. The early portion of the record doubtless includes some twilight. The photomultiplier tube draws a small amount of current even when no light falls on the photocathode, an effort that establishes the minimum illumination to which the tube is sensitive. It is called the 'dark current,' and is indicated

on the graph by the horizontal broken line near the bottom.

"A second 931-A photomultiplier is being added to operate an alarm system during sleeping hours. It will be actuated by light dispersed to the side of the axis opposite the camera location. The characteristic green line at 5,577 angstroms is focused by a small lens on a second slit that excludes all other light from the photomultiplier. For mechanical compactness the beam is folded by a small front-surfaced mirror [*see bottom illustration on page 104*] that can be rotated during initial adjustment to center the 5,577 line on the slit. The output of the photomultiplier is amplified for operating a relay that in turn triggers an alarm when the 5,577-angstrom line reaches a predetermined intensity.

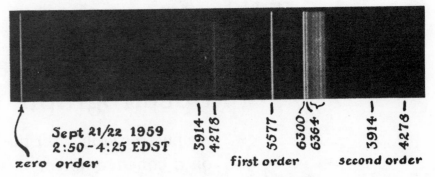

*Spectrum made with the auroral spectrograph*

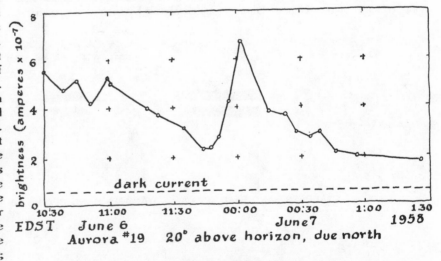

*Graph showing the changing brightness of an aurora*

# 21

# Inexpensive Diffraction-Grating Spectrograph

*A spectrograph with the grating mounted on a concave mirror*

September 1966

Light from any luminous gas or vapor contains in its constituent colors a remarkable amount of information about the source. The colors indicate the chemical elements in the gas or vapor, the proportions of the elements, their atomic structure, their temperature, their motion toward or away from the observer and the strength of the surrounding magnetic field. This information eludes the unaided eye. To obtain the information experimenters use the spectrograph, an instrument that sorts light waves according to length and displays the result as a band of parallel lines resembling a fence of randomly spaced pickets arrayed in the color sequence of the rainbow. During

the century since Gustav Kirchhoff and Robert Bunsen constructed the first practical spectroscope the characteristic pattern of lines emitted by nearly all the naturally occurring chemical elements has been measured and catalogued. Upward of a million lines have been observed but no two elements have ever been found to have a single line in common, although some are very close together.

Few amateurs own spectrographs, because instruments of reasonably good quality cost several hundred dollars. It is now possible, however, for an amateur to build an inexpensive spectrograph. Sam Epstein, chief chemist of the Federated Metals Division of the American

Smelting and Refining Company in Los Angeles, has recently designed an instrument of excellent quality that can be built at home for less than $100. With it experimenters can readily identify approximately 70 chemical elements listed in the periodic table, sometimes even if their presence in a mixture of substances amounts to no more than a few parts per million. In addition the apparatus can be adapted for use with telescopes in analyzing phenomena on the sun, although its size limits its application to permanently mounted telescopes.

Epstein writes: "Physically all spectrographs consist essentially of three elements: a narrow slit, through which

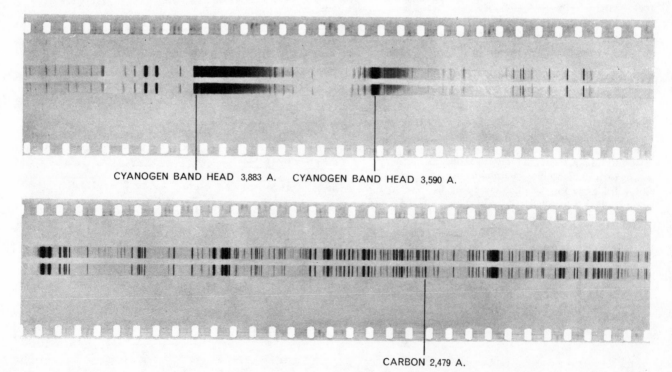

CYANOGEN BAND HEAD 3,883 A.    CYANOGEN BAND HEAD 3,590 A.

CARBON 2,479 A.

*Spectrogram of graphite made with Sam Epstein's homemade spectrograph. Key bands are identified in angstrom units*

passes the light that is to be analyzed; a dispersing element, which may be either a glass prism or a grating that consists of a pattern of closely and uniformly spaced lines ruled on either a transparent or a reflecting surface, and a camera. For a number of years all spectrographs were based on an optical principle first described by Isaac Newton. When a narrow beam of sunlight passes through a glass prism in a darkened room, a pattern of rainbow colors forms on the opposite wall. In the instrument based on this principle a narrow beam is formed by passing light rays through a slit. Diverging rays from the slit are made parallel by a lens; after passing through the prism they are focused on a screen by a second lens. The screen can be replaced by photographic film or the colors can be observed directly with a small telescope. The resulting spectrum consists of a band of adjacent colored lines that are images of the slit.

"Good instruments of this type are difficult to build because they require lenses of high optical quality that focus all colors equally. Moreover, the resolving power, or ability of the instrument to separate the closely spaced lines, increases with the size of the prism, and so does the cost. Glass is opaque to ultraviolet rays—those wavelengths shorter than 3,300 angstrom units constituting the part of the spectrum that is most useful for identifying unknown atoms.

"For these reasons gratings made by ruling lines on the surface of a concave mirror have largely replaced prisms in spectrographs. Gratings used in general-purpose spectrographs may have 30,000 or more evenly spaced rulings cut on a surface of polished metal only an inch wide and two inches long. Such gratings are costly. In recent years, however, methods have been developed for making plastic duplicates of the rulings. These replicas are then mounted on the concave surface of aluminized glass mirrors. As assembled in the instrument the grating receives light either directly from the slit or after reflection by a mirror and focuses the lines of color on a photographic film or plate. All rays are reflected without significant absorption. The resulting spectrum spans some 30 octaves, including the single octave of light.

"An easily constructed spectrograph that can serve as a powerful analytical tool in many fields of experimentation is based on a discovery of the physicist Henry A. Rowland. In the 19th century he observed that if a concave grating, a slit and photographic film are placed on a circle equal in diameter to the radius of curvature of the grating, the diffracted rays come to a focus on the film.

"Unlike the prism, the grating presents the spectrum simultaneously at a number of positions. At one point on the circle the undispersed image of the slit appears. This narrow image is flanked on both sides by a series of spectra, the ends of which may overlap more or less depending on the design of the grating and the angle at which the incoming rays impinge on the grating. These images are known as spectral orders, the undispersed image being designated as the "zeroth" order and the flanking spectra as the first, second and third orders and so on. In general the spectrum of the first order is the shortest and brightest. By cutting the rulings at a certain angle, however, it is possible to construct gratings that reflect most of the light into a particular order. The spacing between the spectral lines increases in proportion to the increased length of the higher orders.

"If the spacing between the rulings of the grating is known, as well as the angle at which incoming light rays fall on the grating, the angular position at which light of any color will come to a focus on the circle can be computed simply. An example can be given for a grating ruled with 15,000 lines per inch. A wave of light 6,300 angstroms in length falling on the grating at an angle of 19 degrees with a line perpendicular to the face of the grating will be diffracted at an angle of 2.9 degrees from the perpendicular and on the same side of it. A wavelength of 2,300 angstroms striking the grating at an angle of 19 degrees will be diffracted at 10.9 degrees on the opposite side of the perpendicular.

"If the 15,000-line grating is supported on the face of a concave mirror with a radius of curvature of 100 centimeters, these two lines, which span 4,000 angstroms, will be separated by a distance of 25.4 centimeters at the plane of the film. That is about 16 angstroms per millimeter. This resolving power is adequate for the analysis of most metallic substances except those that emit a large number of closely spaced lines; examples are iron and the rare-earth elements. A replica grating of this size can be obtained from the Edmund Scientific Co., Barrington, N.J. 08007. The catalogue number is 50,220.

"Begin the construction by drawing on a flat surface a circle with a radius of 53 centimeters. On this 'Rowland circle' locate the exact positions of the slit, grating and camera or film holder as specified in the accompanying illustration [below]. The outline of the

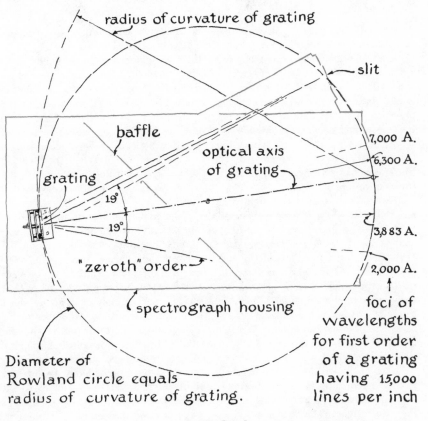

*A Rowland circle*

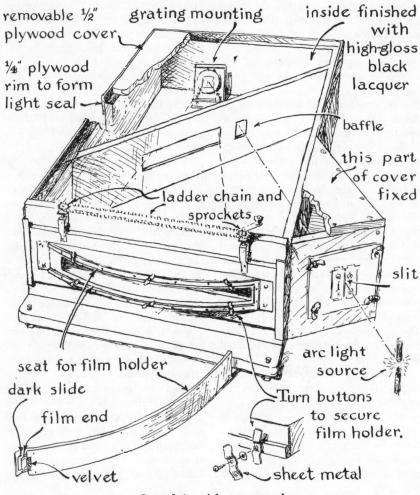

removable ½"
plywood cover

grating mounting

inside finished
with
high-gloss
black
lacquer

¼" plywood
rim to form
light seal

baffle

this part
of cover
fixed

ladder chain and
sprockets

slit

seat for film holder

dark slide

arc light
source

film end

Turn buttons
to secure
film holder.

velvet

sheet metal

*General view of the spectrograph*

spectrograph housing, also shown in the illustration, should be superposed on the circle. The dimensions of the housing are not critical but construction problems will be minimized if they are followed closely. The light baffle not only prevents scattered light from entering the camera and fogging the film but also serves as a support for the top of the housing. All interior parts are painted black to minimize unwanted reflections from the various surfaces.

"The camera consists of a lightproof box that encloses a film holder curved to match the circumference of the Rowland circle [*see illustration on page 110*]. The curved members of the box are wooden arcs of slightly smaller radius than the Rowland circle. This difference in radius compensates for the thickness of the film holder and thus ensures that the emulsion of the film follows precisely the curvature of the Rowland circle.

"Each of the wooden arcs is attached to the camera support by wood screws, and additional support is provided by the steel angles. The machine screws needed to attach the film holder are

fastened in place with epoxy cement. The several brass strips that constitute the film holder are also assembled with epoxy cement. Coat the surfaces as indicated and mount the assembly on the camera by means of the nuts. When the cement has hardened, the holder will have assumed permanently the shape of the Rowland circle. Do not neglect to install black felt or velvet for blocking light at the indicated points where the removable slides provide access to the film.

"The camera slides up and down on ways. Light is admitted to the film through the thin horizontal slot in the mask of sheet metal. Altering the vertical position of the camera in its ways enables the experimenter to record several spectrograms on a single sheet of 35-millimeter film. An indicator on the racking mechanism marks the position of the film with sufficient accuracy so that successive exposures can be separated by .5 millimeter. An internal shutter, which can consist of a hinged flap of sheet metal, must be installed in order to close the entrance slit through which light is admitted to the spectro-

graph. The shutter can be operated either electromagnetically or by a mechanical shaft that extends through the side of the housing.

"The slit consists of a pair of safety-razor blades [*see illustration on page 111*]. Two slit assemblies should be made, one with a spacing of about 50 microns for use with intense light sources and another of 150 microns for observing flames. To set the 50-micron spacing loosen the screws that clamp the razor blades, slip a sheet of notepaper between the cutting edges, press the edges snugly against the paper and tighten the screws. Use a stack of three or four sheets for adjusting the wider slit.

"As I have indicated, spectral lines shorter than 4,000 angstroms are of most interest in spectroscopic analysis. They must be photographed because the eye is insensitive to this part of the spectrum. The Eastman Kodak Company manufactures two special emulsions for spectrographic work: Spectrum Analysis No. 1, SA 421–1, and Spectrum Analysis No. 3, Sp 421–1. Both are sensitive down to about 2,200 angstroms. The upper limit of the No. 1 emulsion is about 4,500 angstroms (deep blue); that of No. 3, about 5,300 angstroms (green). Panchromatic film must be used for recording the complete visible spectrum. The special spectrographic emulsions are sold in 100-foot rolls that cost about $12 each. Single rolls can be bought from Spex Industries, Inc., 3880 Park Avenue, Metuchen, N.J. 08841.

"Unless the experimenter insists on achieving the best possible results in the extreme portion of the ultraviolet range, ordinary black-and-white emulsions can be used. The usual precautions should be taken to avoid accidental exposure when loading the film holder. The films should be developed for high contrast. I use Kodak D-19 developer. The negative is used for analysis. No positive print is required. When making an exposure, pull the opaque slide out just far enough to clear the film aperture of the holder and remember to push it back again before removing the holder from the camera.

"Atoms emit bright-line spectra only when they are excited in the gas or vapor state. At relatively low temperature, however, such as the temperature of an ordinary gas flame, this applies to only a few elements, including potassium, lithium, calcium, barium, strontium and copper.

"Some 25 additional metals can be added to the list by substituting a flame of acetylene and compressed air for natural or manufactured gas. The results of flame excitation can be improved by preparing the specimens in the form of a chloride salt solution. The solution is sprayed into the flame with an atomizer during the exposure.

"The most widely used sources of excitation are electric arcs and sparks. The arc is struck between rods of highly purified graphite specially manufactured for spectrographic work and is sustained by a direct current of about eight amperes. Electrodes of spectrographic grade can be bought from the Ultra Carbon Corporation, Post Office Box 747, Bay City, Mich. 48706. An inexpensive power supply for operation from a 110-volt, 60-cycle power line can be constructed by connecting as a full-wave rectifier four silicon diodes of 10-ampere capacity that are designed for 200-volt operation. These can be bought from most dealers in radio supplies. A resistor must be connected in series with one side of the power line to limit the current. It can consist of a replacement unit of the type used in radiant heaters. A 750-watt unit will do, as will a group of seven 100-watt incandescent lamps connected in parallel.

"The carbon electrodes of the arc must be supported in vertical alignment by a pair of insulated jaws that can be moved together for striking the arc and then separated as soon as the carbon rods become hot enough to maintain an arc. A rack-and-pinion mechanism is the preferable means of moving the jaws. Small used arc lamps easily modified for supporting the carbons vertically can be obtained from dealers in theatrical supplies. The arc should be fully enclosed except for an exit window that will allow a beam about an inch in diameter to fall on the slit of the spectrograph. An experimenter must protect his eyes by wearing welder's goggles when the arc is in operation.

"The material to be analyzed is applied to the tip of the lower carbon rod, which is made the cathode. An axial hole half the diameter of the carbon rod can be drilled to a depth of about an eighth of an inch in the tip of the cathode for admitting powdered specimens. A shallow cut by a saw can also be used. Fluid specimens can be applied to the rod by dipping the tip of the cathode into the solution and letting it dry. Highly volatile specimens should be mixed with powdered graphite to retard the rate at which they evaporate in the arc. Some highly refractory specimens that volatilize at a rate too slow for analysis can be mixed with ammonium chloride. Heat decomposes this salt, which then carries the unknown materi-

al into the arc.

"To adjust the spectrograph for operation remove the slit from the instrument and substitute a slit about one millimeter wide made by placing two strips of masking tape over the slit aperture of the housing. Cut a strip of white paper 32 centimeters long and 10 centimeters wide. This strip is used as a temporary screen. Bisect it lengthwise with a heavy black line and make two vertical lines to indicate the ends of the spectrum aperture. Tape the strip to the inside of the camera section. Center the horizontal line with respect to the opening. Darken the room and direct a flashlight beam into the slit. Adjust the grating so that the resulting spectrum is centered vertically and positioned at the long-wavelength end of the screen. Remove the paper.

"Place a few large crystals of rock salt on a metal screen above the flame of a gas burner and focus the resulting yellow light on the slit with a magnify-

ing lens. The burner should be about 18 inches from the slit. Tape a piece of translucent wax paper three inches wide across the camera opening so that it covers the 5,893-angstrom point, which is about 2.6 centimeters from the long-wavelength end of the opening. A broad yellow line will appear on the screen. Adjust the grating until this line occupies the 5,893-angstrom position.

"Remove the paper and install the permanent slit. Load the camera and place it in position on the instrument. Rack the film holder to either the top or the bottom. Charge the bottom electrode with iron filings. Put the arc about 12 inches from the slit, aligned so that the beam transmitted by the slit floods the grating. Make exposures by striking the arc and opening the shutter. The correct exposure is the one that results in the greatest range of line density. It must be determined experimentally.

"The line images may be fuzzy. If so, the instrument should be focused. Set

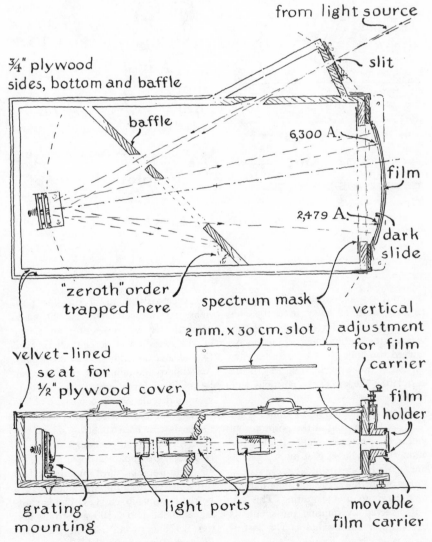

*Plan and elevation views of Epstein's spectrograph*

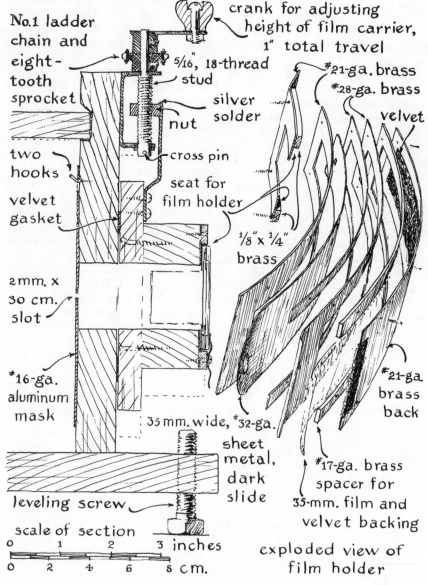

No.1 ladder chain and eight-tooth sprocket

crank for adjusting height of film carrier, 1" total travel

5/16", 18-thread stud

silver solder

nut

cross pin

two hooks

seat for film holder

velvet gasket

#21-ga. brass
#28-ga. brass

velvet

1/8" x 1/4" brass

2 mm. x 30 cm. slot

#16-ga. aluminum mask

#21-ga. brass back

35 mm. wide, #32-ga. sheet metal, dark slide

#17-ga. brass spacer for 35-mm. film and velvet backing

leveling screw

scale of section

0    1    2    3 inches
0    2    4    6    8 cm.

exploded view of film holder

*Details of the film holder*

the grating at one of its limits of travel, either toward or away from the film. Then make a series of exposures, advancing the grating in equal increments toward the limit of its travel after each exposure. Keep a written record for correlating the successive positions with the exposures.

"After developing and examining the film, set the grating at the position that yields the sharpest image. If close examination discloses that each spectral line actually consists of a closely spaced group of fine lines, the slit is not parallel to the rulings of the grating. This can be corrected by rotating the slit in its oversized mounting holes. The instrument is now ready for use.

"Analysis of spectrograms requires the

precise measurement of the line positions. The measurements can be made conveniently by projecting an image of the lines on a screen along with that of a scale calibrated in angstroms. A standard 35-millimeter projector can be modified for projecting the spectrogram and scale simultaneously.

"The standard used for calibrating the scale can be a spectrogram of carbon. Strike an arc between a pair of clean spectroscopic carbons and make an exposure. The most prominent features on the resulting spectrogram will be the two cyanogen (CN) bands starting at 3,883 and 3,590 angstroms and the 2,479-angstrom carbon line [*see illustration on page 106*].

"Lay the film on a flat surface and

measure the distance between the 2,479-angstrom line and the head of the band beginning at 3,883 angstroms. Assume that it is 88 millimeters. If so, each millimeter represents 16 angstroms $(3,883 - 2,479/88 = 16)$. One hundred angstroms will occupy 100/16, or 6.25, millimeters, and a scale that is 25 centimeters long will span 4,000 angstroms.

"Draw the scale on white cardboard, divide it into 40 increments of 100 angstroms each, photograph the drawing and make a positive transparency with an enlarger. The final image should be precisely 25 centimeters long. Draw an auxiliary scale 12.5 centimeters long and divide it into 20 equal increments. The ratio of this scale to the one previously drawn is 20:1; when the scale previously drawn is magnified 20 diameters by projection, the distance between adjacent graduations is 12.5 centimeters and spans 100 angstroms. Line positions can be measured to within one angstrom by placing the auxiliary scale between a pair of projected graduations.

"To determine positions of the lines emitted by an unknown chemical element, make a spectrogram of carbon and on the same film a spectrogram of the unknown element. Insert the film in its compartment in the holder and move the spectrogram with respect to the scale until the 3,883-angstrom point of the scale coincides in position with the 3,883-angstrom cyanogen band head. The distance between the projector and the screen must of course be adjusted for a magnification of 20 diameters.

"Using the auxiliary scale at the screen, measure in angstroms the spectral positions of the lines you believe to be those emitted by the unknown element. Always look for the most prominent lines of an element when trying to establish its presence in the sample. Extensive tables of wavelengths can be found in *The Handbook of Chemistry and Physics*. At least two lines of an element must be detected before it can be considered as being definitely present. Not even the roughest estimate of its presence can be made, however, unless the spectrogram can be compared with one of a similar type of sample material containing a known amount of the element in question.

"Imperfections in a grating can produce 'ghost,' or spurious, lines. Ghosts are easily spotted. They always occur in telltale pairs spaced symmetrically on opposite sides of the parent, or true, line, which is usually of high intensity. Ignore the ghosts.

"The spectrograph can be used for the analysis of many different types of material such as glasses, soils, ceramic materials and ashed substances of all kinds. It is particularly applicable for detecting fractions in concentrations of 1 percent to .0001 percent and even less in some cases. Amateur prospectors, for example, can readily check rocks for metals in amounts far below those required for ordinary chemical or blow-pipe analysis."

## Note on the Grating

*November 1966*

Sam Epstein of Los Angeles, who designed the spectrograph described in this department in September, calls attention to the fact that the instrument was designed for a diffraction grating with a focal length of 53 centimeters. If a grating of differing focal length is substituted, appropriate changes must be made in the dimensions of the instrument.

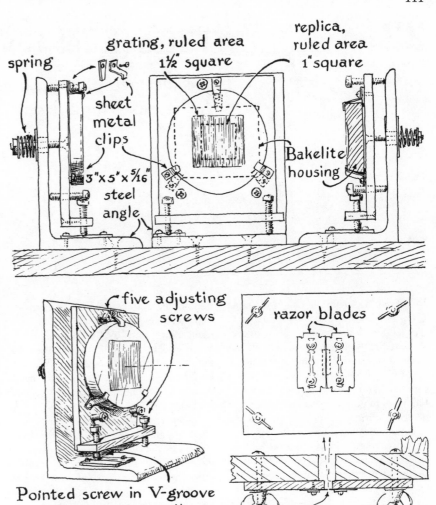

Pointed screw in V-groove locates grating laterally.

slit 50 µ wide

*Arrangement of the grating and the slit*

# 22

# Ultraviolet Spectrograph

*A spectrograph with a quartz prism*
*for work in the ultraviolet*

October 1968

Beginning where the blue of the rainbow fades into invisibility is a broad but normally unseen band of radiation that carries much of the story of the sun. The band is the ultraviolet region of the solar spectrum. To explore it and comparable spectra emitted by all luminous gases and vapors one needs an ultraviolet spectrograph, an instrument that separates relatively short electromagnetic waves according to length and records them on photographic film as a series of parallel lines that vary in spacing and intensity.

The pattern of the recorded lines is determined both by the kind of atoms in luminous gas that emit and absorb radiant energy and by the atoms' energy levels and velocities. A unique spectrographic pattern is associated with each kind of atom. The pattern is influenced in minor but significant ways by the atom's environment.

In terms of the frequency at which electromagnetic waves vibrate, the solar spectrum spans more than 30 octaves, including the single octave of visible light. Spectroscopists have measured, identified and tabulated some three million spectral lines. With this information and an instrument for recording ultraviolet spectra amateurs can identify the metals in ores, learn the relative velocities of particles in glowing gases and vapors, measure colors precisely and perform many related experiments. An inexpensive ultraviolet spectroscope has been built at home by Roger Hayward, the retired architect and optical designer who illustrates this department. Hayward writes:

"Figuratively, at least, I have spent some of my most delightful hobby hours chasing rainbows. It all grew out of a pre-high-school desire to build an instrument that was good enough to show separately the two closely spaced yellow lines that appear in a spectroscope when a sheet of asbestos paper soaked in salt brine is held in the flame of a Bunsen burner. The instrument I planned was to consist of a slit through which light from the Bunsen burner would be admitted to a lens. The lens would make the diverging rays parallel and direct them into a glass prism that, like a drop of rain, would split the light into its constituent colors. The emerging colors would enter a second lens, through which I hoped to see a band of colored lines, each an image of the slit. In particular I hoped that the yellow emission lines of sodium in the salt would appear.

"A glass prism that dangled from the chandelier in our dining room was liberated to do the splitting of colors. A lens from a magnifying glass was appropriated for making the light rays parallel. A lens from another magnifying glass was used as the telescope lens for examining the spectrum. The slit was made from a tin can.

"When these parts had been assembled in a wooden box, I fired up the burner and had a look. I could make out

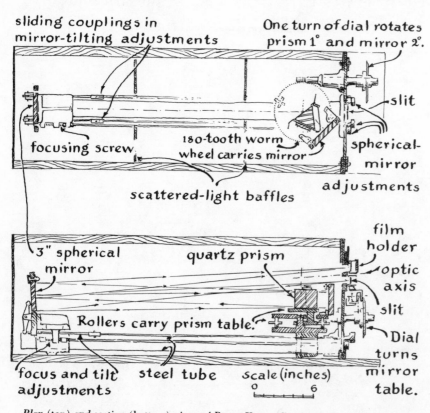

*Plan (top) and section (bottom) views of Roger Hayward's ultraviolet spectrograph*

a single yellow line with fuzzy edges, but no amount of adjustment made it split into the sodium pair. Finally I gave up in disgust, partly at my own ineptness and partly because of the inadequacy of the illustrations in my father's old high school physics text.

"The memory of that frustrating experience goaded me for nearly 30 years. Eventually I decided to try again and also to practice the art of making informative illustrations. In the meantime my interest in spectroscopy had expanded. I learned that the spectroscopic information of most interest lies outside the visible spectrum, much of it in the ultraviolet region to which glass is relatively opaque.

"Two basic schemes have been devised for dispersing ultraviolet radiation with minimum loss. In one of them the rays are reflected from a polished metallic surface, such as aluminum, that has been ruled with closely spaced parallel lines. This forms a diffraction grating, which causes reflected waves to interfere with one another selectively so that the angle at which they emerge from the rulings increases in proportion to their length. At the time I made my spectrograph diffraction gratings were not available at a price I could afford.

"The second scheme is to use a prism cut from a substance that is transparent to ultraviolet radiation. Quartz is such a substance. Fused quartz would work, but at that time it could not be made in large blocks of the necessary optical quality. The resolution of a spectrograph—the ability of the instrument to separate closely spaced spectral lines—increases with the size of the prism. I wanted a prism with at least two inches on a face, which meant I had to cut the prism from a natural crystal of quartz.

"This requirement introduced another complication. Crystalline quartz is an optically active substance. It polarizes light, meaning that it rotates the plane in which the light waves vibrate. It is also birefringent: unless the light travels along a path that parallels the optic axis of the crystal an entering ray is split. It emerges from the other side of the crystal as two rays that produce a double image.

"The effect of birefringence can be minimized by cutting the prism from the quartz in a direction such that the rays travel parallel to the optic axis of the crystal. The effect of polarization can be minimized by placing behind the crystal a mirror that reflects the rays back through the quartz. Waves that are rotated in one direction during their forward transit through the prism are untwisted by the same amount during the return trip.

"In 1938 I bought a two-pound quartz crystal from the Brazilian Importing Company in New York for $5. (No doubt it would cost much more today.) The facets came to a point at one end, which indicated that the optic axis was parallel to the length of the piece.

"Quartz is difficult to cut, so I took the crystal to a shop that had a diamond saw. The technician simply held the crystal by hand and ran it against the blade, using kerosene as a coolant. After cutting out a block roughly in the form of an equilateral triangle we examined the piece critically. The corners were not filled out. It was clear that if we took thin slices off the sides, the prism would present a better appearance. So the technician took off the two slices by hand. As I recall I paid $10 for the job.

"We had just guessed at the angle of the apex. It turned out to be approximately 63 degrees. Prisms of commercial spectrographs are usually made with 60-degree angles, at which the loss of light by reflection from the faces is minimal. The extra loss at 63 degrees is trivial.

"The faces of the roughly cut blank were ground flat and smooth against a surface made of hexagonal ceramic tiles stuck to an eight-inch steel disk with hot pitch. For the grinding compound I used

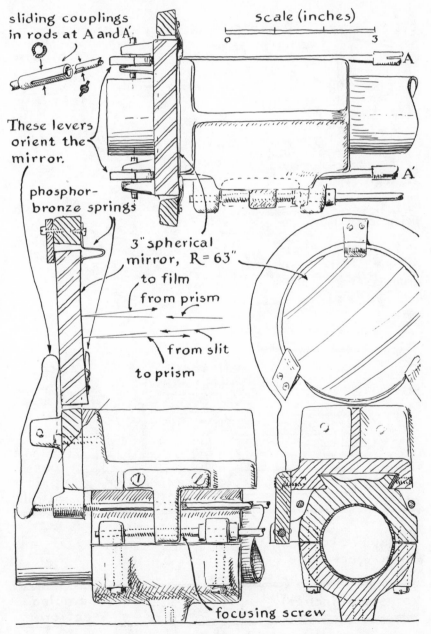

*Details of the mounting of the spherical mirror of the spectrograph*

a slurry of Carborundum grains in water. The rough grinding was begun with No. 80 grit and continued with successively finer grits through No. 600, as explained in *Amateur Telescope Making: Book One,* by Albert G. Ingalls, which is published by SCIENTIFIC AMERICAN. I held the prism by hand and ground it back and forth across the ceramic surface, much as amateurs make telescope mirrors. To polish the faces I covered the ceramic blocks with pitch and used a slurry of rouge in water. The prism was supported by a wooden block during polishing.

"I cannot remember why I elected to grind and polish the prism by this unorthodox procedure instead of mounting it conventionally in a plastic disk along with pieces of quartz to fill out the circular array. I suppose I was just impatient to see the finished product. Anyway, I knew that if the faces did not come out flat, I could always mount the piece conventionally and make the necessary corrections. One of the secrets of my success in handwork of this kind is that I have cold hands. The heat transmitted to the piece by my fingers did not distort the surfaces. People with warm, fleshy hands usually have difficulty getting good optical figures when they attempt handwork, but skinny hands like mine make out well.

"Having finished the prism, I made four mirrors. One had a flat surface for reflecting rays through the prism, one was spherical and two were cylindrical.

The cylindrical mirrors collect light from the source and direct it through the slit of the instrument. The spherical mirror receives converging rays from the slit, transforms them into parallel rays, directs the parallel rays through the prism, receives the dispersed rays and reflects them to a focus on a strip of photographic film located immediately above the slit [*see illustration on page 112*].

"The glasses were ground, polished and figured by techniques somewhat similar to those used for making telescope mirrors. A commercial shop applied reflecting films of aluminum to the polished glasses. Incidentally, disks of cast iron that are used as weights on barbells can be made into handy flat surfaces on which to grind mirrors. They are generally available from dealers in sports accessories. The hole in the center is not much bother. I had the face of a disk ground flat on a Blanchard grinder by a local machine shop.

"The instrument is housed in a light-proof pine box with Masonite ends. A tube of 16-gauge steel 1½ inches in diameter forms the backbone of the mechanism for supporting and adjusting the position of the optical parts. One end of the tube slides into a bronze casting, locked by clamps, that forms a table for supporting the spherical mirror.

"The table has a sliding top for altering the distance between the mirror and the other optical elements of the instrument [*see illustration on preceding page*]. A threaded hole in a lug that extends downward on one side of the sliding tabletop engages a screw that can be turned by a knob on the front of the instrument. This control adjusts the focus of the spectral lines on the film.

"Two other knobs and similar screw mechanisms operate a pair of rocker arms that bear against the rear edges of the spherical mirror, which is supported in an annular cell by spring clips. By manipulating the screws I can adjust the angular position of the mirror. The bottom of the casting includes a foot that is fastened to the bottom of the box with a screw.

"The design of the mechanism for mounting and positioning the prism and the flat mirror is somewhat more elaborate. The prism must be rotated to bring the various regions of the spectrum into view and to center desired spectral lines on the photographic film. To accomplish the rotation certain optical properties of the prism must be taken into account in the design of the mounting fixture.

"A ray of light that traverses a prism is bent where it enters the prism and bent

*Details of the mounting of the prism assembly*

still more in the same direction when it emerges into the air. The magnitude of the total angle through which a ray is bent depends on the nature of the crystal and also on the angle between the ray and the face of the prism through which it enters. The ray is bent least when the angles between the ray and the front and rear faces of the prism are equal. Then the path of the ray is symmetrical with respect to the faces of the prism. When the prism is so positioned, it is said to operate at minimum deviation.

"Short waves are bent progressively more than long ones. The waves are dispersed according to their length. Some transparent substances disperse waves of differing lengths more than other substances do. It turns out that the dispersion of quartz is relatively low. For this reason I decided to use a 60-degree prism instead of one of 30 degrees, as is customary in instruments of this type.

"This decision made it imperative that I operate the prism at minimum deviation because, as mentioned, quartz is birefringent. If a quartz prism is operated at other than minimum deviation, the oppositely polarized rays do not cancel completely, and two images appear in the focal plane. Of course, not all rays can be made to pass through the prism at minimum deviation because some waves are bent more than others, depending on their length. Even so, the prism can be placed so that the rays of most interest traverse it on a path that closely approaches minimum deviation.

"To achieve this condition it is essential when scanning the spectrum to rotate the flat mirror at exactly twice the rate of the prism. A system of differential gears would do the trick, but I was worried about the smoothness of gear motion. Gear teeth tend to introduce periodic accelerations.

"I decided to generate the desired motion by rotating a disk on a set of three conical rollers linked with yokes. One revolution of the disk would cause a half-revolution of the roller assembly [see illustration on opposite page]. The movement of the rollers is restrained by a pair of circular grooves. One groove is in the bottom of the disk and the other is in a circular baseplate.

"The disk, which is a flat worm gear, carries the mirror. A hole through the gear admits the foot of a prism table to the roller assembly, to which the table is attached by screws. The width of the hole restricts the rotation of the disk to 30 degrees, but in use the table turns only 26 degrees. Two of the rollers are yoked to the prism table. The third roller

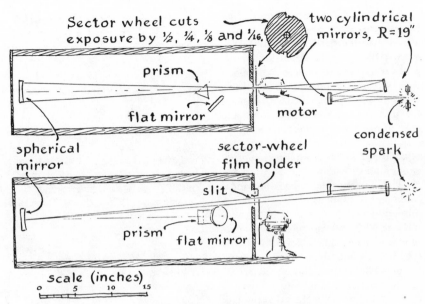

Optical path of the spectroscope

acts as an idler. A worm engages the teeth of the gear and is rotated by a shaft that terminates in a calibrated dial on the front of the instrument.

"Most of the metal parts are castings of red brass. They could have been made from strip, plate and rod stock of other metals. I undertook the project just after I had written a chapter on molding and casting for the book Procedures in Experimental Physics, by John Strong and others. It therefore seemed to me that I should design the apparatus to be built with castings made from my own patterns. There are 18 patterns all told, one for a wavelength drum that I later discarded. The bill from the foundry, as I recall, amounted to about $20. I sometimes wonder what the same job would cost today.

"The control panel consists of a brass plate that covers an opening in the front of the cabinet. It fits loosely in a groove in the edge of the Masonite front. In turn the plate supports the shaft bearings and control knobs, the slit assembly and the film carrier. When the spectrograph was completed, I learned that the spherical or autocollimating mirror could have been equipped with less fancy controls. The position of the mirror has not been altered since its initial adjustment more than 25 years ago!

"Ports for the slit assembly and the film carrier are identical. They are separated by a narrow dovetail strip. The optic axis of the system passes through the center of the strip. For this reason the slit and the film carrier can be interchanged.

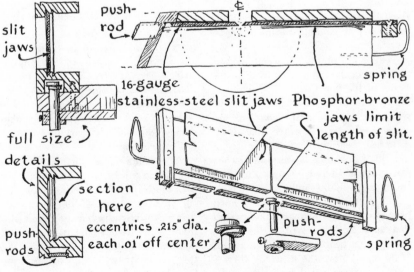

Details of the slit assembly

"Access to the optic axis can be had by removing the dovetail strip, which is attached to the panel by four machine screws. By removing the strip and the prism it is possible to direct light along the optic axis by means of a half-silvered mirror. One can then look through the half-silvered mirror and adjust the positions of the flat and spherical mirrors so that light is returned along the optic axis. The mirrors are then in proper orientation.

"The dial that rotates the flat mirror is calibrated in minutes of angular arc. A setscrew locks it to its shaft in the zero position when the flat and spherical mirrors have been aligned. Each full turn of the minutes dial advances an associated degrees dial 1/30 of a turn.

"The motion is transmitted from the minutes dial to the degrees dial through a Geneva movement and a pair of 1 : 3 reduction gears. This feature was an afterthought. I could have used a Veeder counter to record degrees, of course, but I had fun filing out the Geneva wheel by hand. I undertook the job mostly to see if I could do it. The cup-shaped degrees dial and the large flat minutes dial are made of Lucite.

"The film carrier was designed to take pieces of cut roll film, Type No. 120, 2¼ inches wide and 2⅛ inches long. Strips 3/4 inch wide can also be used. It is easier to make the film carrier lightproof if the dark slide that protects the emulsion when the carrier is removed from the instrument cannot be fully removed. I fastened a limiting stop to the inner end of the slide; it prevents the slide's removal. The slide was made from sheet brass 1/32 inch thick. The frame of the carrier is milled from stock brass strip one inch wide and 1/2 inch thick. The rim of the cover, which includes a flat spring for pressing the film in place, is also made of sheet brass and is silver-soldered to the top plate. It could be made of other metal.

"The body of the slit assembly was milled from the same stock used for the frame of the film carrier. The width of the slit should be adjustable so that the width of the spectral lines and the amount of light that enters the instrument can be controlled. The center of the slit should not be displaced when the width is altered, because the spectral lines would be similarly displaced.

"The slit of my instrument is formed by two strips of 16-gauge stainless steel that slide in V grooves of the body. Springs that bear against the outer ends of the strips move the strips together. A pair of eccentrics that bear against the

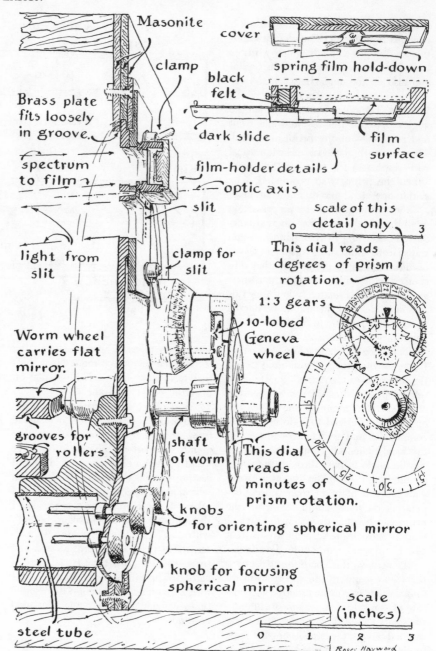

*Arrangement of the spectrograph's control panel*

inner ends of a pair of pushrods, which are coupled in turn to the ends of the strips, force the strips apart [*see bottom illustration on preceding page*]. A mechanism designed to force the strips together, instead of separating them, might damage the inner ends, which are honed to sharp edges.

"The frame of the slit assembly also supports a pair of strips made of phosphor bronze, the inner ends of which are cut at an angle of 45 degrees. These strips can be adjusted to admit rays through any part of the slit. In one position, for example, they may admit rays from an unknown source to the upper half of a piece of film. After the unknown

spectrum has been recorded the position of the strips can be readjusted for recording a known spectrum on the lower half of the film as a comparison. The position of the diagonal strips can be similarly changed for registering a series of exposures of various time intervals.

"The most precise and convenient method of determining the wavelength of an unknown spectral line is to compare its position with that of a line of known wavelength. The spectrum of iron, which is particularly rich in lines that have been accurately measured and tabulated, is used routinely as the comparison spectrum. I generate emission from iron by means of a spark gap.

"The gap consists of a pair of cold-rolled steel electrodes 1/4 inch in diameter; they are sharpened to edges 1/32 inch wide and 1/4 inch long. The rods are supported in alignment by an electrically insulating fixture of plastic so that the inner edges are parallel and spaced 1/32 inch apart. Alternating current is applied to the electrodes by the smallest available neon-sign transformer. The one I use develops about 2,000 volts and perhaps 20 milliamperes.

"I connect a high-voltage capacitor across the gap. I made the capacitor by sandwiching sheets of aluminum foil between glass plates and connecting alternate sheets to the respective electrodes of the spark gap. Odd-numbered foils were connected to one electrode, even-numbered foils to the other.

"The glass plates are about one millimeter thick and of the quality used for supporting photographic emulsion. They are five inches long and 2¼ inches wide. The sheets of foil are 4¼ inches long and 1½ inches wide.

"A 350-micromicrofarad capacitor of the high-voltage type used in amateur radio transmitters would serve nicely. Without the capacitor in the circuit the spark is faint and bluish and radiates many spectral lines that arise from gases and chemical radicals in the atmosphere. With the capacitor it becomes a brilliant bluish white and the light is not contaminated by the atmosphere.

"For operation the instrument requires a final adjustment. The prism must be mounted on its table and oriented to the position of minimum deviation. Two accessories are needed: a source of monochromatic light and an eyepiece that includes a pair of cross hairs. For the source I use a mercury lamp; a sodium flame could be substituted. The eyepiece is trained on the center of the field normally occupied by the photographic film. The mechanism for adjusting the eyepiece is a mounting plate that slides into the grooves that hold the film carrier.

"Light from the mercury lamp is focused on the slit. With the prism table stationary I rotate the prism by hand, thereby moving the spectral line at a wavelength of 5,461 angstrom units (the most brilliant green line of mercury) into the field of view. As the prism is rotated still more in the same direction the green line advances at a constantly decreasing rate and finally comes to rest. At this point the prism is set for minimum deviation. If the prism is rotated still more in the same direction, the green line resumes its movement but in the reverse direction. The prism is lightly clamped to its table in the position of minimum deviation by means of a knurled screw.

"To record the iron spectrum I replace the eyepiece with the loaded film carrier, open the slit to about .004 inch and flood it with light from the spark gap. The time of exposure must be determined experimentally. Several exposure intervals can be tried by periodically advancing one of the two diagonal jaws to increase the length of the spectral lines by increments. A second spectrum, such as that of mercury, can then be recorded adjacent to the spectrum of iron. The brilliant green line, which is known to be located at 5,461 angstroms, is easily identified on the developed film and can serve as a starting point for identifying the iron lines by reference to published data.

"The spectral lines of iron are numerous and closely spaced. Their positions are most easily measured by means of a traveling microscope, which consists of a mechanical carriage with a calibrated micrometer screw for transporting a low-power microscope across the recorded spectrum at precisely measured intervals of length. A traveling microscope I made was described in this department in August, 1954. The accompanying photograph [right] shows a small length of the ultraviolet spectrum of iron that was made with my spectrograph and analyzed with the traveling microscope."

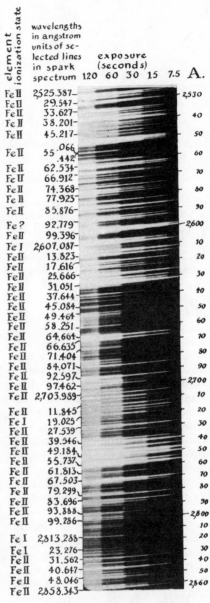

| element | ionization state | wavelengths in angstrom units of selected lines in spark spectrum | | A. |
|---|---|---|---|---|
| Fe | II | 2525.387 | | |
| Fe | II | 29.547 | | |
| Fe | II | 33.627 | | 40 |
| Fe | II | 38.201 | | |
| Fe | II | 45.217 | | 50 |
| Fe | II | 55.066 | | 60 |
| | | .442 | | |
| Fe | II | 62.534 | | 70 |
| Fe | II | 66.912 | | |
| Fe | II | 74.368 | | 80 |
| Fe | II | 77.925 | | |
| Fe | II | 85.876 | | 90 |
| Fe | ? | 92.779 | | 2,600 |
| Fe | II | 99.396 | | |
| Fe | I | 2,607.087 | | 10 |
| Fe | II | 13.823 | | 20 |
| Fe | II | 17.616 | | |
| Fe | II | 25.666 | | 30 |
| Fe | II | 31.051 | | 40 |
| Fe | II | 37.644 | | |
| Fe | II | 45.084 | | 50 |
| Fe | II | 49.464 | | 60 |
| Fe | II | 58.251 | | |
| Fe | II | 64.661 | | 70 |
| Fe | II | 66.635 | | 80 |
| Fe | II | 71.404 | | |
| Fe | II | 84.071 | | 90 |
| Fe | II | 92.597 | | 2,700 |
| Fe | II | 97.462 | | |
| Fe | II | 2,703.989 | | 10 |
| Fe | II | 11.845 | | 20 |
| Fe | I | 19.025 | | 30 |
| Fe | II | 27.539 | | 40 |
| Fe | II | 39.546 | | |
| Fe | II | 49.184 | | 50 |
| Fe | II | 55.737 | | 60 |
| Fe | II | 61.813 | | 70 |
| Fe | II | 67.503 | | |
| Fe | II | 79.299 | | 80 |
| Fe | II | 83.696 | | 90 |
| Fe | II | 93.888 | | 2,800 |
| Fe | II | 99.286 | | |
| Fe | I | 2,813.288 | | 10 |
| Fe | I | 23.276 | | 20 |
| Fe | II | 31.562 | | 30 |
| Fe | II | 40.647 | | 40 |
| Fe | II | 48.046 | | 50 |
| Fe | II | 2,858.343 | | 2,860 |

exposure (seconds) 120 60 30 15 7.5

*A small portion of the iron spectrum*

# 23

# Inexpensive Spectrophotometer

*A photocell to measure the intensity of color transmitted by a liquid*

*May 1968*

Color can serve as a powerful clue to the identity, nature and even the behavior of many substances if the observer can recognize a characteristic shade and perceive small differences of hue. Unfortunately color is difficult to judge by eye and even more difficult to specify precisely in terms of hue. For example, a popular instruction for applying silver to glass when making a mirror calls for adding ammonia to a solution of silver nitrate until the mixture becomes the color of "weak tea." How yellow is weak tea? I learned to recognize the desired shade by mixing, observing, testing and discarding several quarts of costly chemicals until at last a brightly silvered mirror emerged from a solution of the correct color.

Time and money would have been saved if I had owned a spectrophotometer, which is an instrument that measures colors and mixtures of colors in terms of the wavelength of light transmitted by the specimen and also records the intensity of the colors in terms of the percentage of light that is transmitted. Until a few years ago spectrophotometers cost more than I could afford; those of the highest performance still do. The advent of inexpensive electronic and optical parts, however, has made it possible for anyone who is reasonably handy to build a serviceable spectrophotometer at home. One simple design that can be assembled for less than $75 is described by R. C. Dennison of Westmont, N.J. He writes:

"The spectrophotometer can be one of the most useful instruments in an amateur's shop, particularly for the analysis of chemicals. I use mine for determining properties as diverse as the color of glasses and plastics, the transmission of light by neutral-density filters and semi-silvered mirrors, the percentage of chlorine and other substances in water, the kind of metals that may be present in specimens of rock and the composition of alloys. Essentially the instrument disperses light that is transmitted by the specimen into the rainbow hues of the visible spectrum and measures the intensity of the emerging colors one by one.

"The physical scheme of the instrument is simple. Diverging rays from an incandescent lamp pass through a thin mechanical slit, known as the entrance slit, and through a collimating lens that makes the rays parallel [*see top illustration on opposite page*]. The parallel beam passes through the specimen, where certain colors may be fully or partially absorbed, depending on the nature of the specimen. Colors that remain in the beam enter a prism that disperses them into the orderly array of the spectrum.

"The spectrum is focused by a second lens, called the telescope lens, as a band of rainbow colors on an opaque white screen that is perforated with the exit slit. This slit transmits one narrow band of color or another, depending on its position in relation to the spectrum. The transmitted rays fall on a photoelectric cell and induce in it an electric current that varies in magnitude with the intensity of the colored light. The current is measured by a microammeter.

"The amount of light that reaches the specimen is controlled by the position of a wedge-shaped diaphragm that in effect determines the length of the entrance slit. The photocell, the exit slit, the telescope lens and the prism are assembled on a carriage of sheet metal that is attached at the prism end to a vertical shaft. By turning the shaft the operator can move the exit slit across the spectrum in order to select a desired color. A dial fixed to the shaft indicates the position of the slit in terms of the wavelength of light. The microammeter is calibrated to indicate the percentage of light transmitted by the specimen. All parts are housed in a cabinet equipped with a light-tight lid for shielding the photocell from room light when measurements are made.

"The parts are assembled on the bottom and one side of a steel radio chassis (Bud CB-643) that is 17 inches long, 13 inches wide and four inches deep. A second chassis of the same size is hinged to the first as a dust cover and light shield. The lamphouse is made of cookie-sheet aluminum and is 2¾ inches long, 2⅛ inches wide and 2½ inches high. A hole 5/8 inch in diameter in one wall of the box is partly closed by a pair of double-edged razor blades spaced .0045 inch apart to form the vertical entrance slit. A socket that fits a General Electric No. 93 incandescent lamp is mounted on the wall opposite the slit. The lamp is installed with its filament in the vertical plane. The housing is ventilated by a one-inch hole in the chassis and several quarter-inch holes in the top. A baffle of sheet aluminum inside the lamphouse near the top prevents the escape of stray light.

"The collimating lens is mounted on a rectangle of sheet aluminum 2⅜ inches square that contains in the center an aperture 5/8 inch in diameter. Two 1/8-inch tabs cut from the upper corners of the rectangle are bent over as supports for the upper edge of the lens. The lower edge is supported by a tab of aluminum attached to the rectangle by a machine screw.

"The assembly is mounted on a bracket of sheet aluminum 2½ inches wide and 2¾ inches high that is perforated with a centered hole one inch wide. The bracket is attached to the chassis by screws passing through slots in the foot that enable the collimating lens to be moved toward or away from the lamphouse when the lens is focused. The optical axis of all elements of the optical train is 1½ inches above the base.

"The carriage, which includes the photocell, the exit slit, the telescope lens and the prism, is made of sheet brass 1/16 inch thick, 2⅛ inches wide and nine inches long. One edge is bent up 5/16 inch to provide stiffness. The photocell housing is similar to the one for the lamp but consists of a Bud CU-3000 A Minibox. One wall contains the vertical exit slit, made with razor blades spaced .003 inch apart. They face the photocell. The photocell (RCA Type 7117) was designed for automatically dimming automobile headlights. I bought one at an automobile junkyard for a fraction of the list price.

"The cover of the photocell housing consists of a rectangle of Bakelite 1/8 inch thick to which an Amphenol socket (Type 77MIP11) is attached by screws. Nine 33-kilohm resistors that supply voltage to the photocell are soldered directly to the lugs of the socket. A thin sheet of Bakelite, supported by standoff pillars, is mounted over the lug side of the socket to prevent accidental contact with the high voltage.

"The inverted photocell projects downward into the box. The socket must be oriented so that the photocathode faces the exit slit. (Point the keyway of the socket toward the exit slit.) The completed assembly is attached to the carriage arm by machine screws in the position illustrated. Apply a strip of flat white paint to the razor blades and the front of the housing.

"A 1/4-inch vertical shaft is attached by machine screws to the end of the carriage arm opposite the photocell by means of a flange in the form of a brass gear that happened to be on hand [see bottom illustration at right]. Any equivalent flange would serve as well. The distant end of the carriage is supported by a wheel that rides on top of the chassis. The wheel was made by soldering a short hollow rivet into a 3/4-inch washer. The axle of the wheel consists of a machine screw that enters the rear of the photocell housing near the base on the center line of the carriage.

"The vertical shaft is supported and driven by a modified worm gear from a surplus gunsight. An equivalent mech-

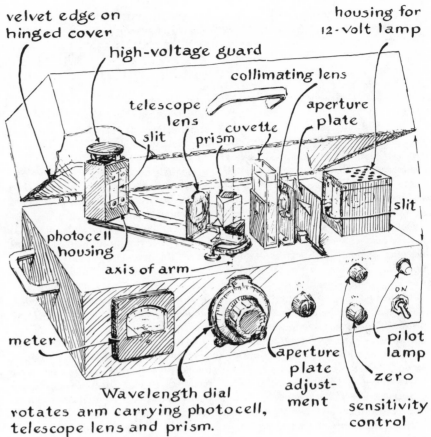

R. C. Dennison's spectrophotometer

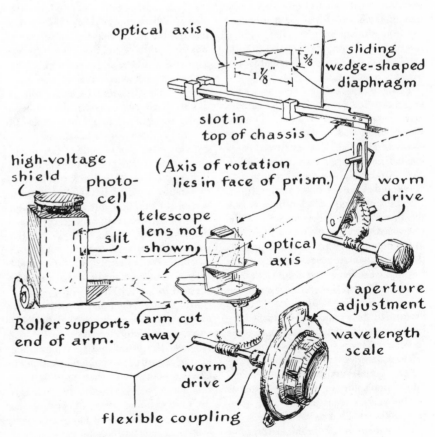

Details of dial and diaphragm mechanisms

anism could be constructed with a Millen Type 10000A worm-gear drive of the kind designed for rotating the tuning capacitors of radios. This unit is available from dealers in radio supplies. It is made in gear ratios of 16 to one and 48 to one. A gear of the latter ratio should be used.

"The horizontal shaft of the worm gear extends through the front of the chassis. It should be equipped with a planetary dial that has a drive ratio of eight to one (Lafayette Radio Electronics stock No. 99 H 6029). When this dial is turned through 240 angular degrees, the carriage sweeps the exit slit completely across the spectrum. The dial is designed for only 180 degrees of rotation, but the plastic stops that limit the rotation can be sawed off by disassembling the unit.

"A small table of sheet metal was improvised to support the prism on the optical axis of the instrument above the end of the vertical shaft. The prism (Edmund Scientific Co. stock No. 30,143) is held in place on the table by a clip made of spring brass. The telescope lens is attached to the carriage arm by a fixture similar to the one that supports the collimating lens.

"The amount of light that reaches the specimen is controlled by a triangular diaphragm cut into a plate of brass that moves across the path of the rays emerging from the entrance slit. One edge of the plate is bent at right angles to form a foot 1/8 inch wide. This foot is soldered to a rectangular brass bar, of about the same width and thickness, six inches long. The bar moves in slotted guides made from two small blocks of Bakelite that are attached to the chassis by machine screws. The bar is driven by a mechanical linkage consisting of a Millen No. 10,012 right-angle drive, a bell crank and a slotted brass arm.

"The right-angle drive is mounted on the inner face of the front wall of the chassis. An arm of sheet brass, soldered to the outer end of the rod that supports the triangular aperture, projects down through a slot in the chassis. A slot about 1/2 inch long in the arm engages a bell crank that is driven by the right-angle drive. The bell crank, when turned by the right-angle drive, advances or retards the triangular aperture. By advancing or retarding the aperture the microammeter can be adjusted to full-scale deflection without altering the current that is normally present in the photocell when the photocathode is in darkness.

"Specimens are inserted in the light path at a point between the collimating lens and the prism. A fixture improvised from sheet metal is attached to the chassis at this point for supporting a rectangular glass box that is called either a cuvette or an absorption cell. Fluid specimens are placed in the cuvette for measurement.

"Cuvettes are priced at $15 and up because they must be made of reasonably flat, well-annealed glass. Glass of the quality used for photographic plates is adequate and can be bought from distributors of photographic supplies. The inside dimensions of the cuvette should be about 75 millimeters, 35 millimeters and 10 millimeters. The ends, sides and bottom piece can be cut with a glass cutter of the wheel type and assembled with epoxy cement. Solid specimens such as colored glasses, filters and semitransparent mirrors are inserted for measurement in the position normally occupied by the cuvette.

"I followed conventional techniques when building the electronic portion of the instrument. Small components were mounted on stiff sheets of perforated plastic known as Vector board. The transistors specified in the accompanying schematic illustration [*opposite page*] were used because they were on hand; they do not necessarily represent either the best or the least costly design. The 2N1702 transistor is mounted on the chassis, which acts as a heat sink, and is insulated from the chassis by a thin mica washer supplied by the manufacturer. The 1N1204RA diodes are mounted directly on the chassis, which again acts as the heat sink. (Incidentally, the anodes of diodes that bear the suffix 'RA' connect to the mounting stud and do not need to be insulated from the chassis.)

"The conventional scale of the microammeter was replaced by one calibrated in intervals from 0 to 100 for indicating the transmission of light in percent. It was also calibrated in units of density, according to the relation: density equals the logarithm, to the base 10, of the ratio 100 divided by the transmission in percent. For example, at 50 percent transmission the density of the specimen is equal to $\log_{10} 100/50$, or .3.

"In order to align and focus the optical system and calibrate the dial that controls the position of the exit slit I first removed the prism and the collimating lens of the fully assembled instrument. The lamphouse was then positioned to center the wedge of light that emerges from the entrance slit on the optical axis of the instrument. The lamphouse was locked in this position by its mounting screws.

"The simple plano-convex collimating lens (1¾ inches in diameter with a focal length of three inches) is mounted with its plane side toward the lamphouse at a point that causes the diverging rays from the slit to become parallel after they have passed through the lens. To locate this position I first focused a pair of binoculars on an object about a mile away. The binocular was then positioned so that the objective lens of one half of the instrument was on the optical axis of the spectrophotometer and faced the collimating lens, looking toward the entrance slit. (A sheet of white paper can be placed between the lamp and the slit to reduce the intensity of the light.) The position of the collimating lens was now adjusted until a sharp image of the slit appeared in the binocular. Any small, low-power telescope can be substituted for the binocular. When the collimating lens was focused, it was locked to the chassis by tightening the mounting screws.

"The telescope lens is assembled in its holder with its plane side facing the exit slit. Again, with a small telescope focused on infinity, adjust the position of the telescope lens on the carriage until a sharp image of the exit slit appears. (Light the slit in front with the beam of a 35-millimeter projector.) Lock the telescope lens in this position.

"Release the setscrew on the rear hub of the dial that drives the carriage. Mount the prism on its table, light the incandescent lamp and slowly rotate the prism back and forth until the spectrum appears on the white surface of the exit slit. Turn the prism back and forth on its table and observe that at one position the angular deflection of the spectrum is at a minimum. The prism is now set at the angle of minimum deviation. Turn the shaft of the worm-gear drive and shift the position of the prism by trial and error until at minimum deviation the center of the yellow band of the spectrum falls on the exit slit. Rotate the shaft to move the slit to the red end of the spectrum, position the dial at the limit of its excursion and lock it to the shaft.

"The dial must now be calibrated to indicate wavelength. If possible, borrow a set of narrow-band interference filters that transmit light of known wavelength. With the exit slit at the red end of the spectrum, insert the interference filter of longest wavelength in the cuvette holder at right angles to the light beam and rotate the dial for maximum photocurrent.

"Record the arbitrary indication of the dial and repeat the procedure for each

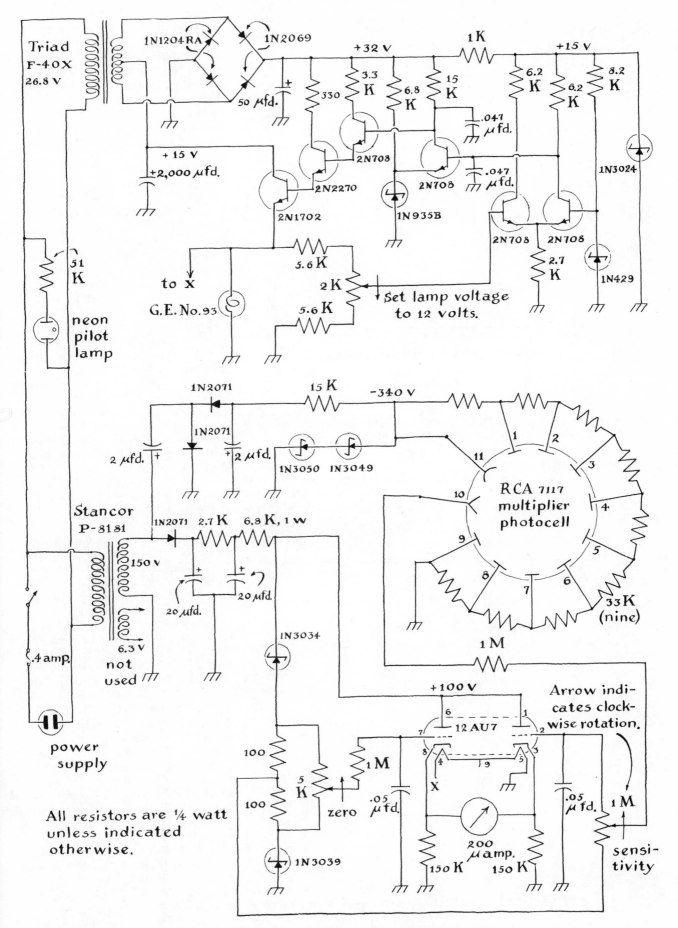

Schematic circuit diagram for the spectrophotometer

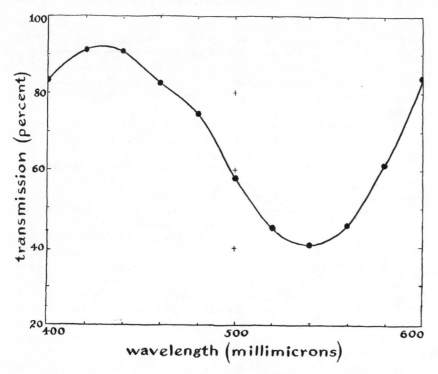

*Spectral response of potassium permanganate solution*

filter. On linear graph paper plot the arbitrary indications of the dial against the corresponding wavelengths. From these data make a scale for the dial calibrated in millimicrons. The scale will be crowded at the red end.

"If interference filters are not available, the instrument can be calibrated

with reasonable accuracy by means of didymium glass. A didymium filter accompanied by a curve of spectral transmittance can be bought from the Arthur H. Thomas Company, P.O. Box 779, Philadelphia, Pa. 19105. The item is listed as didymium filter No. 9104-N20 and costs less than $2. The transmittance

curve supplied with the filter displays nine dips and peaks between 400 and 700 millimicrons. Mount the filter and make a series of readings in which the dips and peaks are correlated with arbitrary dial readings. Convert the arbitrary readings to wavelengths by referring to the calibration curve supplied with the filter and make a corresponding wavelength scale for the dial.

"When the photocurrent of the spectrophotometer is plotted against wavelength, the resulting graph takes the form of a bell-shaped curve that peaks at approximately 550 millimicrons and drops to about 5 percent of the maximum reading at 380 and 660 millimicrons. The graph depicts the intrinsic response ($I$) of the instrument.

"Intrinsic response must be known before an unknown color can be determined. For maximum accuracy of measurement the intrinsic response should be redetermined prior to measuring each unknown specimen. For example, to measure the spectral response of a piece of colored glass, find the wavelength at which the photocurrent is maximum. Remove the specimen and adjust the intensity of the light (by altering the position of the wedge-shaped diaphragm) until the pointer of the meter swings to full scale (the 100 percent indication). Turn the wavelength dial to its limit at the red end of the spectrum. Replace the specimen. Record the meter indication at this wavelength and designate the response $R$. Remove the specimen and designate the resulting response $I$. Make similar pairs of readings at intervals of five or 10 millimicrons across the spectrum to the limit at the blue end.

"When the readings of $R$ and $I$ drop below 20 percent of full-scale meter deflection, as they will doubtless do at the red and blue ends of the spectrum, the instrument will lose some accuracy. This loss can be compensated for by increasing the sensitivity of the instrument. The sensitivity can be increased by turning up the sensitivity control until the limit of usable gain is reached.

"The transmittance ($T$) of the specimen is equal at each interval of wavelength to $R$ divided by $I$. Calculate the transmittance at each interval of wavelength and from the tabulated computations prepare a graph of the spectral response. The response of unknown solutions is also measured. The intrinsic response of solutions is determined by replacing at each interval of wavelength the cuvette containing the specimen with an identical cuvette that contains only solvent.

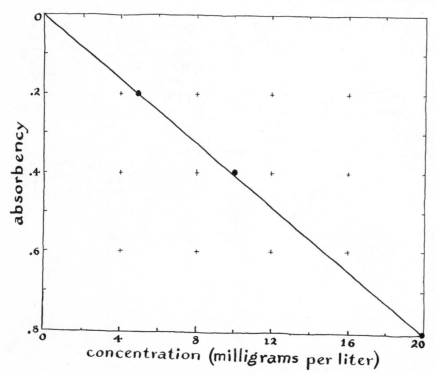

*Calibration graph of spectrophotometer for permanganate ion*

"Experience in the use of the instrument and confidence in the reliability of the measurements can be gained easily by making graphs of the spectral responses of Wratten filters and comparing the results with graphs supplied by the manufacturer. These inexpensive filters and their graphs can be bought from dealers in photographic supplies. Simple experiments for beginners also include the measurement of various dyes, food colors and other colored solutions [see "The Amateur Scientist," SCIENTIFIC AMERICAN; February, 1965].

"A more advanced experiment that demonstrates the usefulness of the spectrophotometer as an analytical instrument consists of the analysis of steel for the presence of manganese. The procedure is based on the magenta color that appears when a solution of manganous salts is oxidized. The amount of manganese in the steel is calculated by comparing the absorbency of the colored solution with the absorbency of a standard solution of potassium permanganate. (Absorbency, like density, is equal to the logarithm, base 10, of 100 divided by the transmission in percent.)

"To prepare the standard solution dissolve 72 milligrams of potassium permanganate in water and dilute the solution to 250 milliliters. In this step and all following procedures use only distilled water and reagent-grade chemicals. Weights and volumes must be accurately determined.

"Transfer four milliliters of the standard solution to a clean container and dilute to 40 milliliters. As thus diluted the solution contains 10 milligrams of manganese per liter. Transfer the diluted solution to a cuvette and make a graph of the spectral response [see top illustration on page 122]. Maximum absorbency occurs at 540 millimicrons. My instrument indicated a transmission at this wavelength of 40.5 percent, from which the absorbency was calculated to be .393. $(100/40.5 = 2.47.$ $Log_{10}$ $2.47 = .393.)$ From the stock solution make similar dilutions that contain five milligrams and 20 milligrams of manganese per liter and tabulate the absorbencies. A calibration curve for use in the subsequent analysis can now be drawn by plotting the absorbencies of these three measurements against concentration in milligrams [see bottom illustration on page 122].

"Saw a small piece of steel, weighing about 200 milligrams, from a bar or rod. Place the sample in a 100-milliliter volumetric flask and add five milliliters of water and five milliliters of nitric acid. Warm the solution until the sample has dissolved. While the solution is warm add sodium bismuthate until a slight excess remains. Dilute to 100 milliliters.

"Transfer 10 milliliters of this solution to a clean vessel and dilute to 100 milliliters. Transfer a specimen of the latter solution to a cuvette. Measure the transmittance and calculate the absorbency. Determine the concentration of manganese by referring to the calibration chart. Assume, for example, that the concentration turns out to be 1.5 milligrams of manganese per liter. Before the dilution the concentration was 15 milligrams per liter. The volume of the original (undiluted) solution was 100 milliliters. Therefore it contained 1.5 milligrams of manganese. The specimen of steel weighed 200 milligrams. Hence the steel contains $1.5/200 \times 100$, or .75, percent manganese.

"Another interesting experiment involves a test for cobalt. To make it you will need the following materials: (1) a few grams of sodium pyrophosphate, which can be made by fusing disodium hydrogen phosphate in a crucible; (2) a 60 percent solution of ammonium thiocyanate, made by dissolving 30 grams of the salt in water and diluting to 50 milliliters; (3) acetone; (4) a standard cobalt solution containing 100 milligrams of cobalt per liter, made by dissolving 49.36 milligrams of cobalt nitrate hexahydrate in water and diluting to 100 milliliters, and (5) another cobalt salt, such as cobalt chloride.

"Transfer 10 milliliters of the standard cobalt solution to a graduated cylinder and add 1/2 gram of sodium pyrophosphate. The sodium pyrophosphate prevents any iron that may be present from discoloring the solution. Add 2.5 milliliters of 60 percent ammonium thiocyanate and mix. Dilute to 25 milliliters with acetone and mix. The clear solution will turn blue.

"Place a specimen of the colored solution in the cuvette, measure the transmission and plot the spectral response. Determine the absorbency at the wavelength of maximum absorption and, by serial dilution and subsequent measurement, tabulate data for plotting a calibration graph, as in the analysis of steel. Check the results by measuring the percentage of cobalt in a solution of cobalt chloride by weight.

"Finally, determine the amount of cobalt in an alloy, such as Alnico. Wrap a small Alnico magnet in cloth and, with a hammer and chisel, break off a few fragments. Weigh a specimen of about 200 milligrams. Dissolve the specimen in 10 milliliters of hot nitric acid. Dilute to 100 milliliters. Transfer 10 milliliters of the solution to a clean vessel and add 1/2 gram of sodium pyrophosphate and 2.5 milliliters of 60 percent ammonium thiocyanate. Mix and filter the solution. Dilute the filtrate with an equal volume of acetone. Measure the absorbency and calculate the percent (in weight) of cobalt in the specimen.

"Caution: Most of these chemicals are toxic. Acetone is highly flammable. Avoid contact with the substances. Do not inhale the fumes of reacting mixtures. Be sure to work in a well-ventilated room."

# 24 Recording Spectrophotometer

*A recording spectrophotometer built by a high school student*

*January 1975*

The identity and even the behavior of many substances can be determined by measuring their color with a recording spectrophotometer. The instrument is essential for the precise determination of color because the human perception of color is partly subjective. The instrument measures only the spectral character of the light reflected or transmitted by the specimen. An example of the subjective factor is that a sheet of paper the observer believes to be white is perceived as white whether it is viewed in sunlight, in the red rays of the setting sun or in the light of a yellow flame. Conversely, a fabric of unknown color may appear greenish when it is examined indoors under an incandescent lamp but appear distinctly blue in the light of the northern sky.

The recording spectrophotometer floods a specimen with the constituent colors of the spectrum sequentially, measures with a photocell the amount of light at each hue that the specimen transmits or reflects and with an automatic pen recorder plots a graph of the measured intensity of the light with respect to its wavelength. The graph fully describes the constituent hues that in sum constitute the color. The price of commercial instruments ranges from several hundred to several thousand dollars depending on their resolution, that is, the discreteness with which they plot differences in hue. An instrument that is adequate for many experiments by amateurs and that fully demonstrates the principles of the recording spectrophotometer has been improvised at a cost of about $50 by a Canadian high school student, Sean Johnston (4447 Venables Street, Burnaby, British Columbia, V5C 3A5). He describes his instrument as follows:

"Having built several spectroscopes, I decided last year to make a spectrophotometer that resembled the one described in 'The Amateur Scientist' several years ago. In it rays of light from an incandescent lamp that diverge from a slit are made parallel by a lens, dispersed into their constituent hues by a prism or a diffraction grating and finally made to converge through a second slit one color at a time, depending on the angle of the prism or the grating with respect to the axis of the optical train. Such a device is known as a monochromator. It can separate any color of light from a mixture of colors. If the dispersing element is rotated through an angle of sufficient magnitude, all hues can be made to appear sequentially at the output slit. A photocell on which this output is incident can be wired to generate a voltage that varies in proportion to the intensity of each hue.

"The addition of the photocell light meter converts the monochromator into a spectrophotometer [see "The Amateur Scientist," SCIENTIFIC AMERICAN, May, 1968]. To reduce the cost of materials I substituted a diffraction grating of the reflection type for the prism that served in the instrument described above. I also replaced the vacuum tubes of the photometer with solid-state devices. The reflection grating (No. 50,201) was purchased from the Edmund Scientific Co. (300 Edscorp Building, Barrington, N.J. 08007). The cost of the project was subsequently increased significantly when I added an automatic pen recorder, thus converting the instrument into a recording spectrophotometer. The motor that drives the recording pen was also obtained from the Edmund Scientific Co. (No. 71,702).

"The available literature indicated that the best spectrophotometers split the light from a single source into two parts, a principal beam and a reference beam. Both beams proceed through the monochromator portion of the instrument, in which the light is dispersed into its constituent colors. The color of each beam can be independently altered, however, before the light is dispersed. For example, one might arrange for the principal beam to be intercepted by a chemical solution and the reference beam to be intercepted by the solvent. A photometer can be designed that in effect subtracts the colors of the reference beam from those of the principal beam. The output of a photometer so designed would represent the true colors of the solute.

"Two general schemes have been devised for making the subtraction. One is known as the 'double beam in time' and the other as the 'double beam in space.' In the first scheme the principal beam and the reference beam as they emerge from the monochromator are directed sequentially onto a photocell by an oscillating mirror or an equivalent device. The photocell accordingly generates an alternating current that is amplified to run an electric motor, which operates the recording pen.

"The direction in which the motor runs is determined by the phase of the alternating current; therefore it is determined indirectly by the relative amplitude of the beams. Simultaneously the motor moves an 'optical comb,' or mask, to intercept the reference beam to a certain degree. The interception adjusts the reference beam to an intensity that matches the intensity of the principal beam. The scheme, which is also known as an 'optical-null a.c. servomechanism,' simultaneously and automatically moves the recording pen to the corresponding null position of the graph. The system is highly stable, primarily because alternating-current amplifiers are inherently more stable than direct-current ones.

"Notwithstanding this desirable feature, I resorted to the second scheme, involving the double beam in space, because the instrument is far simpler and cheaper to build. In this arrangement the dispersed color of each beam is incident on a companion photocell. I substituted inexpensive cadmium sulfide photocells for the photomultiplier tubes found in commercial instruments.

"To subtract the output of the reference beam from that of the principal beam I connected the photocells in series to function as adjacent arms in the circuit of a Wheatstone bridge, which has four arms. The other two adjacent arms are two variable resistors [see *upper illustration on page 127*]. The resistance of the photocells varies inversely with the incident illumination. When the incident illumination is equal on both cells, the resistances of the cells are equal. If the variable resistors are then adjusted to be equal, the bridge is balanced and no potential difference exists between the

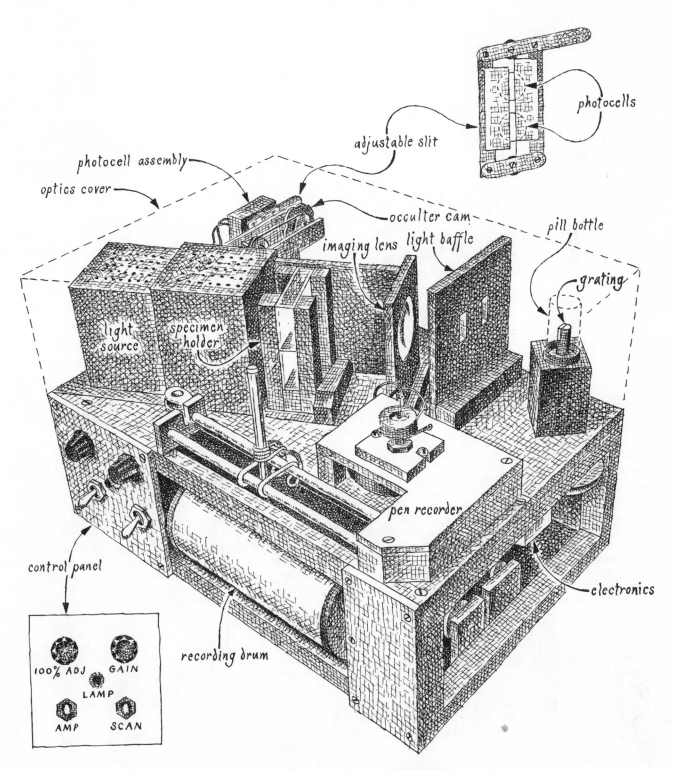

*Sean Johnston's recording spectrophotometer*

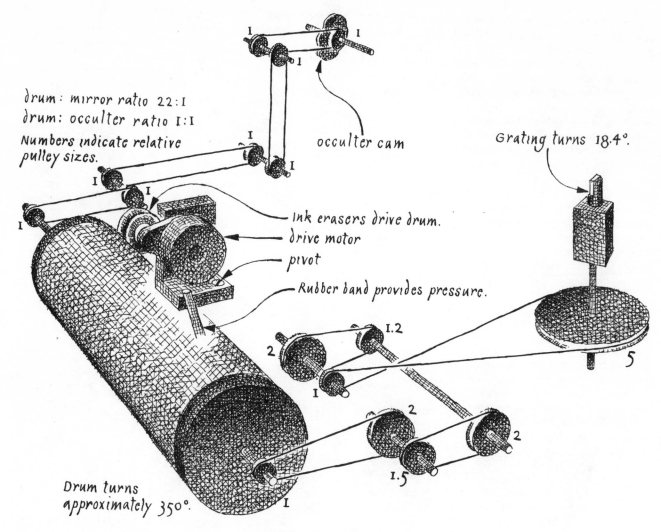

drum: mirror ratio 22:1
drum: occulter ratio 1:1
Numbers indicate relative
pulley sizes.

occulter cam

Grating turns 18.4°.

Ink erasers drive drum.
drive motor
pivot
Rubber band provides pressure.

Drum turns
approximately 350°.

**Arrangement for propelling the drum**

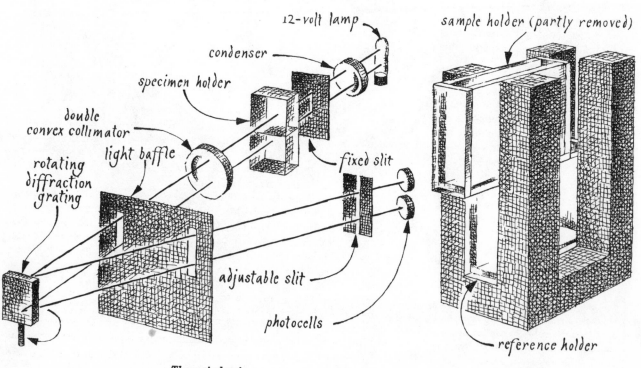

12-volt lamp
condenser
specimen holder
double
convex collimator
light baffle
rotating
diffraction
grating
fixed slit
adjustable slit
photocells

sample holder (partly removed)

reference holder

**The optical train**

**The specimen holder**

junction of the resistors and the junction of the photocells even though the circuit contains a three-volt battery.

"If the intensity of either light beam varies, however, a proportionate potential appears across the now unbalanced bridge and hence across the input terminals of the amplifier. The amplified potential operates the direct-current motor. The shaft of the motor assembly rotates the movable contact of one of the variable resistors in the Wheatstone bridge in the direction required to balance the circuit, thus reducing the input potential of the amplifier to zero. Simultaneously the motor moves a recording pen through a distance that corresponds to the algebraic sum of the light intensity of the two beams.

"The recorder consists of a felt-tipped pen that moves in a straight line parallel to the axis of a rotating drum around which the graph paper is wrapped. A graph is made by rotating the drum through 350 angular degrees. A system of pulleys and belts that is coupled to the drum simultaneously and synchronously rotates the diffraction grating through an angle sufficient to scan the visible spectrum from violet (400 nanometers) to dark red (750 nanometers) [see top illustration on opposite page].

"A second system of pulleys and belts similarly rotates an opaque cam called an occulter. The cam partly intercepts the reference beam as required to compensate for deficiencies in the system, with the result that a graph indicating 100 percent transmission of all colors, as with a water-white specimen, approximates a straight line. The contour of the cam is being determined experimentally and filed by hand. I am still working to improve it.

"The drum consists of two soup cans fastened end to end with epoxy cement and adhesive tape. The assembly is covered by a length of rubber inner tube that is secured at the ends with broad rubber bands. The drum is rotated through frictional contact with a pair of typewriter erasers of the disk type that are fastened with epoxy cement to the shaft of a synchronous motor, which turns at one revolution per minute. Motors of this kind are common in electric clocks.

"The motor is fastened to a supporting bracket by a single screw on which it can pivot in the vertical plane. Firm contact between the rotating typewriter erasers and the rubber band at one end of the drum is maintained both by gravity and by the tension of a pair of rubber bands stretched between the motor and

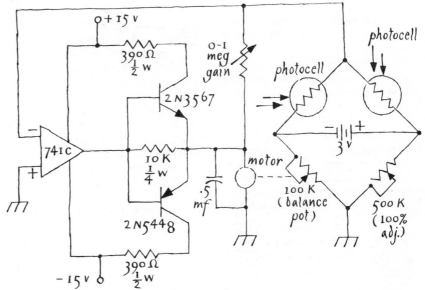

*Electronic circuit of the instrument*

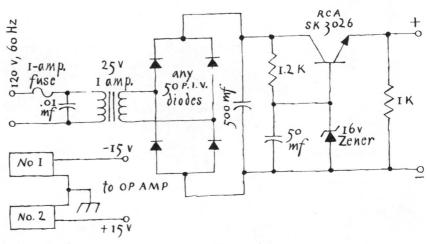

*Circuitry of the power supply*

the base. The pulleys of the drive system came mostly from a Meccano set. The belts are of catgut.

"The specimen is scanned by the full spectrum during approximately one revolution of the drum. Thus the apparatus generates a graph that depicts the spectral response of specimens consisting of an unchanging mixture of colors, such as colored glass or gelatin filters. By disconnecting the belt that couples the drum to the diffraction grating, graphs can be made by plotting changes in intensity of a selected color against time. This effect is frequently associated with chemical reactions. The angle of the diffraction grating can be set by hand to expose such specimens to any desired

hue. When changes in the intensity of the transmitted hue are plotted against the rate at which the drum rotates, they measure the speed of the chemical reaction.

"The most difficult but most interesting part of the apparatus to develop was the pen mechanism. As I have mentioned, the electric power for operating the motor that moves the pen is the amplified potential that appears across the Wheatstone bridge. The signal appears whenever the bridge is out of balance in response to variations in the intensity of the light that falls on the photocells. The amplitude of the signal varies in proportion to the net intensity of the light beams.

"There is a minimum level below which the motor that moves the pen will not respond. Moreover, when balance is restored to the bridge by the operation of the balancing resistor that is coupled mechanically to the motor, the pen does not stop instantly at the point of balance. Inertia causes the mechanism to overshoot.

"For these reasons the graph of a smoothly increasing signal can appear as a series of stairsteps. When the signal current increases to the minimum required to start the motor, the pen moves upward abruptly. Because of the mo-

mentum it continues to move upward briefly after balance is restored. The movement of the pen then stops but the transverse motion of the paper, which is carried by the rotating drum, continues. The result is a graph in the form of a step instead of a smooth line. The solution is to minimize the mass of the moving parts, to minimize friction and to amplify the signal appropriately. By these stratagems the size of the steps can be reduced to the point at which they merge into a continuous line.

"Almost any small, reversible direct-current motor can be installed to operate

the pen. The Edmund No. 71,702 motor that I selected includes a built-in set of reduction gears that turn the output shaft 60 revolutions per minute when the motor is connected to a source of .011 ampere at a potential of three volts. With this power input the motor develops a torque of eight inch-ounces. I increase the force at the pen severalfold by additional speed reduction.

"The shaft of the motor carries a rubber friction roller that I made by pushing a twist drill through a rubber eraser. Turning the assembly at high speed with an electric hand drill, I held a piece of

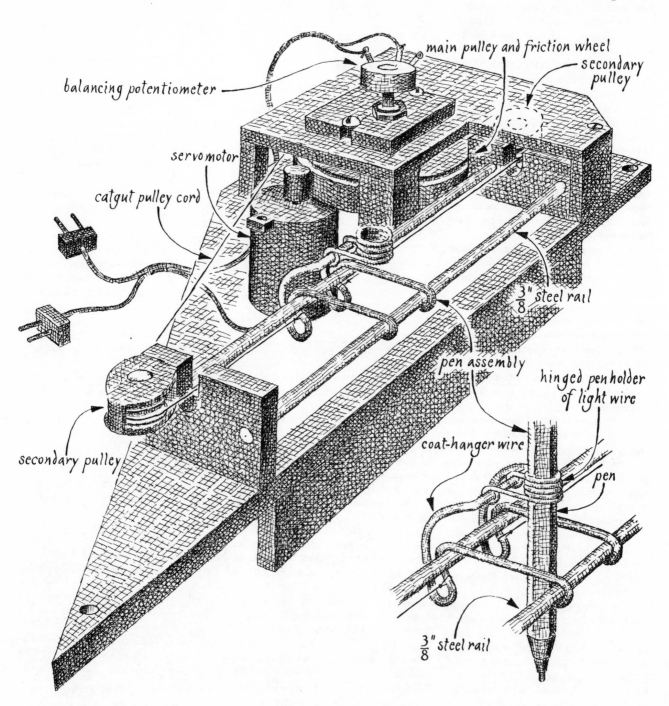

*Details of the pen servomotor*

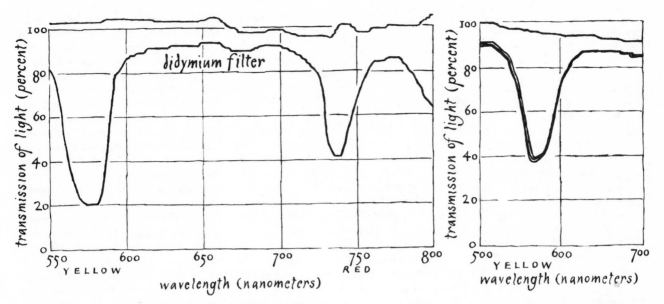

*Transmission graph of didymium filter*

*Three graphs of didymium superposed*

sandpaper against the eraser until the rubber was eroded to a smooth cylinder. The rubber was cemented to the shaft. It presses against the rim of a four-inch plywood disk that is locked on the shaft of a conventional potentiometer. A four-inch pulley is also locked on the shaft. The potentiometer, the motor, the rails and two secondary pulleys are mounted on an independent framework of quarter-inch plywood [*see illustration page 128*]. The rails guide a sliding carriage made of coat-hanger wire. The carriage supports a helical coil of smaller wire that is free to swing in one vertical plane. The helix makes a snug fit with the felt-tipped pen. Its freedom in the vertical plane enables the pen to follow irregularities in the surface of the paper.

"The secondary pulleys were bought at a local hardware store. Half of the block in which each pulley was mounted was cut off with a hacksaw and discarded. The cut face of each block was cemented to the plywood frame with epoxy. One end of the catgut cord that transmits power was tied to the carriage. The other end was then threaded over the pulley at the outer end of the carriage rails, around the motor-driven pulley, around the pulley at the inner end of the rails and then returned to the carriage, to which it was tied. The rails are three-eighth-inch iron rods of the kind sold for hanging window drapes.

"The two power supplies for the amplifiers represent the costliest part of the construction, particularly the step-down transformers that reduce 120 volts to 25 volts. I happened to have materials on hand for making the entire instrument except the electronic components. They cost approximately $40. Doubtless the cost could have been cut in half if I had taken the time to search the surplus market.

"The usual precautions should be observed when the electronic components are assembled. For example, in bending the leads of a solid-state component always grasp them close to the devices to avoid cracking the seals. In making solder joints connect an alligator clip to the leads or grasp them with long-nose pliers to obtain a heat sink. The 2N3567 and 2N5448 transistors or equivalent devices should be provided with heat sinks, which can be of the snap-on type. Incidentally, any transistors can be substituted in this application provided that they are rated at a beta of 40 to 100, a power dissipation of at least .3 watt and a collector-to-base potential of at least 40 volts.

"The optical system consists of a 12-volt incandescent lamp of the kind used in spotlights, a condensing lens for collimating the rays that diverge from the lamp, a pair of fixed entrance slits, a companion pair of specimen holders, a focusing lens, an aperture, a rotatable diffraction grating of the reflecting type, an exit aperture, an adjustable pair of exit slits and a cadmium sulfide photocell for each of the two beams [*see illustration at bottom left on page 126*]. Specimens are inserted into the light beams at an arbitrary point between the entrance slit and the focusing lens.

"The supporting fixture consists of a plywood frame into which cuvettes are

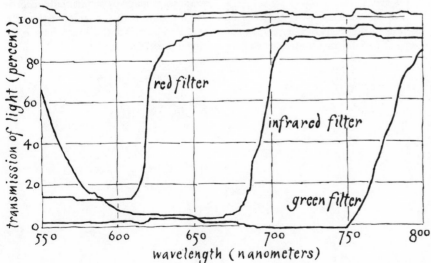

*Graphs of three types of filter*

slid. (A cuvette is a clear, rectangular container of glass or plastic for holding solutions.) In my arrangement the specimen solution is placed in the upper cuvette for interception by the principal beam; the reference solvent is placed in the lower cuvette. Cuvettes can be made by cementing or waxing together appropriately sized rectangles of clear, flat glass of the optical grade used for projector slides. This scheme applies only if the experimenter excludes test solutions that would react with the cement or the glass. I also made a set of small frames that fit the supporting fixture to hold glass or plastic light filters.

"Cadmium sulfide photocells are designed for maximum sensitivity to a specific color. For example, the Clairex Type CL-702 is most responsive to the blue end of the spectrum and is suitable for use with fluorescent light. The sensitivity of the Clairex Type CL-705 is maximum at 550 nanometers and closely matches the spectral response of the human eye. This type is commonly employed for light measurements and is suitable for use with incandescent lamps. The response of the Type CL-703 peaks at 735 nanometers in the deep red portion of the spectrum. It can be used with either incandescent or neon lamps. Types that peak at intermediate wavelengths are also available from mailorder distributors that specialize in electronic supplies, such as Allied Electronics Corp. (2400 West Washington Boulevard, Chicago, Ill. 60612).

"The interception of the reference beam by the occulting cam tends to increase the output of the amplifier just as though the intensity of the principal beam had been increased. In effect the occulting cam acts as a 'phantom specimen' of negative transmittance. Normally the adjustable resistor of the Wheatstone bridge is operated to shift the recording pen to a position on the graph that the experimenter arbitrarily selects

as the point of 100 percent transmission. If the instrument were perfect, and if the specimen were completely clear, the resulting graph would be a straight line at the 100-percent-transmission level. The instrument is far from perfect. For this reason the uncorrected graph of 100 percent transmission is an irregularly undulating curve that extends above and below the 100-percent-transmission line [see illustration top left p. 129]. By altering the contour of the occulting cam experimentally, however, I have succeeded in limiting the excursion to less than 3 percent. By continuing to alter the shape of the cam I hope to reduce the error to less than .5 percent.

"The less than perfect response of the instrument can be traced to a number of obvious sources. For example, the monochromator could be improved by adding a telescope lens between the diffraction grating and the exit slit. The spectral response of the cadmium sulfide photocells is far from linear. Of greatest importance, however, is the relatively low quality of the diffraction grating, particularly with respect to resolution. The stock from which the grating is made was originally developed to split light into its constituent hues for a system of color photography that was later abandoned. I hope to replace the material eventually with a diffraction grating of instrument quality, such as the Edmund No. 41,028.

"Most supports of the optical elements were made from plywood either a quarter or an eighth of an inch thick. The grating support is a block of wood that measures $2 \times 1\frac{1}{2} \times 1\frac{1}{2}$ inches. The grating is fragile. To protect it I included the inverted lid of a plastic pill bottle as part of the mounting assembly. When the grating is not in use, I snap the pill bottle over it.

"For convenience in testing the apparatus during construction and in subsequent maintenance I made each func-

tional element as a removable subassembly. Parts made in this way include the recording drum, the servomotor, the optical train and the electronics system. A panel on the front of the instrument includes on-off switches for the lamp, the drum motor and the electronic system and knobs for adjusting one variable resistor of the Wheatstone bridge and the gain of the amplifier. This latter control is rarely used because I operate the system at maximum gain.

"To begin operating the instrument I attach one edge of a sheet of blank paper to the drum with adhesive tape, rotate the drum by hand to wrap the paper snugly and fasten the remaining edge with another strip of tape. I then switch on the electronic system, including the lamp. Before inserting specimens in the cuvettes I set the diffraction grating by hand so that blue light (450 nanometers) falls on the photocells.

"The adjustable resistor of the Wheatstone bridge is then operated to the point where the pen carriage moves close to the left edge of the graph paper. I insert the pen in the carriage. The drum motor is switched on. The pen traces the 100-percent-transmission line at the top of the graph.

"After one full revolution the drum is stopped, the grating is returned to its initial position and specimens are inserted. The drum is again operated to record the graph. The first of the accompanying graphs [top left on page 129] depicts the characteristic opaqueness of a didymium filter to yellow light. The second graph was made by resetting the diffraction grating three times and recording three traces of the didymium filter on a single sheet of paper to demonstrate the repeatability, or precision, of the instrument. The remaining set of graphs was made by sequentially recording the transmission characteristics of a red, an infrared and a green filter on a single sheet of paper."

# Spectroheliograph

# 25

*A spectroheliograph to observe details
on the disk of the sun*

*April 1958*

Ten years ago a young coal miner in West Virginia sent a letter to this department which began: "Those who helped make the amateur-telescope-making books possible have caused me to live two years of my life in complete contentment." The letter went on to tell how its author, Walter J. Semerau, who now lives in Kenmore, N.Y., had constructed a six-inch reflecting telescope. In the intervening years Semerau has had a remarkable career both in amateur optics and in his daily work. He left coal mining to become an electrician, then an instrument-maker, then a laboratory technician and finally an engineer. Meantime, as regular readers of this department have learned, his six-inch telescope has been succeeded by a whole galaxy of bigger and better instruments, including a 12½-inch reflector complete with a coronagraph and a spectrograph. Semerau now informs us that his telescope mounting supports a new apparatus which has long been the dream of amateur telescope makers; a spectroheliograph of the Hale type. This instrument provides him with a view of the sun rarely enjoyed by laymen.

"Although the sun is a fairly stable body of gas," writes Semerau, "it is neither as amorphous nor as placid as the casual viewer might suppose. Immense clouds of ionized hydrogen, calcium and other substances thrown up from the interior account for features as distinct as the earth's oceans and land masses. Although each of these features emits light of unique color and intensity which distinguishes it from its surroundings, they are lost in the white glare of the sun as it is seen by the naked eye. To see the details clearly the observer is obliged to examine the sun in light of a single color.

"One might suppose offhand that the details could be brought into view by looking at the sun through a filter of colored glass. This stratagem would fail because colored glass, however deeply it is stained, transmits a broad band of colors, just as a radio set of poor selectivity permits several broadcasting stations to be heard at the same time. One must use a filter with an extremely narrow 'pass-band.'

"Such a device was hit upon about a century ago in India by the French astronomer Pierre Janssen. Janssen was using a spectrograph equipped with two slits to observe a total eclipse of the sun. The image of the sun's edge was focused on one slit. Rays proceeding through the slit were spread out by the prism into the familiar ribbon of spectral lines. Janssen was examining one of the lines through the second slit—the dark red line characteristic of glowing hydrogen—when he saw a tongue of flame standing out from the solar edge. Opening the slit brought more and more of the prominence into view until the width of the slit exceeded that of the red line. At this point the image became blurred. To examine slit-shaped portions of the solar disk in other colors Janssen simply shifted the viewing slit to other lines of the spectrum.

"Some 40 years later George Ellery Hale and Henri Deslandres independently devised a method of using the double-slit spectrograph to make photographs of the whole solar disk. The two slits were simply coupled mechanically so that they could be moved as a unit. When the entrance slit is swept across the sun's image, the exit slit keeps in step with the similarly moving spectral line of any selected color. Solar features emitting light of that color are focused on a photographic plate and build up a composite image that resembles the scanned image of a television picture. The device, called the spectroheliograph, was only a step away from the spectrohelioscope, which presents the composite image to the eye. To make a spectroheliograph into a spectrohelioscope one simply arranges for the slits to oscillate across the sun's disk at a rate of 20 or more sweeps per second and substitutes an eyepiece for the photographic plate.

"Not many spectroheliographs have been built by amateurs because of the difficulty of procuring the element which disperses white light into its constituent colors. This may be either a glass prism or a diffraction grating. Prisms large enough for the job are hard to make, and no amateur has succeeded in ruling a diffraction grating of the required fineness and precision. In recent years, however, the Bausch & Lomb Optical Company has introduced an excellent and relatively inexpensive 'replica' grating: a plastic cast of an original grating. A replica grating two inches square with 15,000 rulings per inch costs no more than a set of good golf clubs. With it the amateur can build a spectrograph of exceptional performance and equip this basic instrument with an accessory for making spectrohelioscope observations. [For a description of Semerau's spectrograph see "The Amateur Scientist"; September, 1956.]

"Although it is possible to fit out a spectroheliograph for mechanical scanning, the arrangement is bulky, difficult to maintain and a remarkably effective generator of unwanted vibrations. For these reasons I adopted the optical-scanning system devised by Hale in 1924. The conventional rocker arm which carries the entrance and exit slits in the mechanical system is replaced by a pair of rotating glass cubes, or Anderson prisms. The image of the sun is focused on the fixed entrance-slit of the spectrograph through one prism. The image of the similarly fixed exit-slit is focused on the plateholder (or on the focal plane of the eyepiece) through the second prism. As the prisms rotate, refracted light sweeps past the slits as though the slits had been moved across the rays mechanically. The prisms are mounted on the ends of a shaft which turns on ball bearings; the unit can be assembled on the mounting of even a small telescope without introducing perceptible vibration.

"The spectroheliograph assembly consists of (1) a main housing to which the moving parts are attached and (2) a tube for the eyepiece, reflex mirror and 35-millimeter camera [see drawing on page 134]. The unit is relatively light, compact and simple in construction. It weighs 10½ pounds complete with eyepiece and camera, and measures 15 inches over all. An adapter makes the assembly interchangeable with the plateholder of the spectrograph, which is

mounted beside the telescope. The bearings of the equatorial mounting have enough friction to offset the added weight of the unit; thus no change is required in the counterbalance when the spectroheliograph is used.

"Each prism is clamped between a flange at the end of the shaft and a metal disk held in place by through bolts. Rubber sheeting between the glass and metal protects the prisms against excessive mechanical strain. Center to center the prisms are 3.625 inches apart, the distance between the entrance and exit slits of the spectrograph. The flange supporting the outer prism is grooved for an 'O' ring belt through which the rotating assembly is coupled to a miniature direct-current motor. The facets of the prisms must be adjusted to lie in a common plane or the image will flutter when the unit is started up.

"Parallel rays entering my telescope come to a focus at a distance of 62.5 inches from the 12.5-inch objective mirror, a focal ratio of $f/5$. The focal ratio of the spectrograph is $f/23$. To feed the spectrograph with light from the telescope a set of negative achromatic lenses was introduced into the optical path be-

tween the objective and the spectrograph. This compensates for the difference between the focal ratios of the two instruments. A pair of front-silvered optical flats was mounted at the upper end of the telescope to receive rays reflected from the objective and bend them 180 degrees into the spectrograph.

"Incoming rays pass through one rotating Anderson prism, scan the entrance slit and diverge to an eight-inch spherical mirror at the opposite end of the spectrograph tube, where they are reflected as parallel rays to the diffraction grating at the other end of the tube. Here the white light is dispersed into its component colors and returned to the spherical mirror, which brings the resulting spectrum to a focus in the plane of the exit slit [see drawing on next page]. The exit slit may be adjusted to match the width of any spectral line. The most useful lines are the red 'alpha' line emitted by glowing hydrogen and the 'H' and 'K' calcium lines in the violet region of the spectrum. Light transmitted by the exit slit proceeds through the second Anderson prism, the scanning action of which, together with a final lens assembly, re-

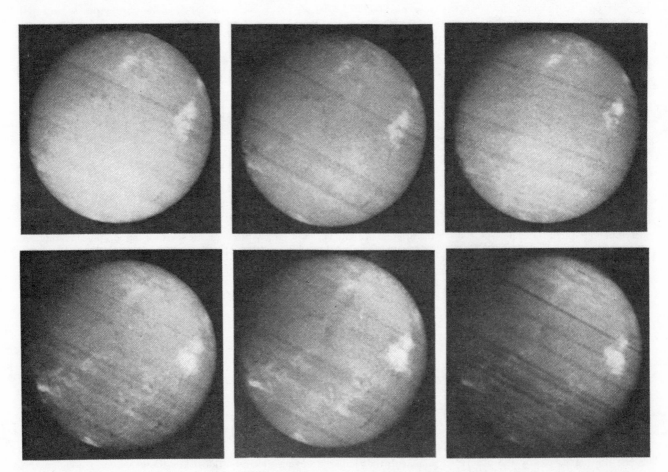

*These spectroheliograms were made on October 20, 1957 at 14:15, 14:18, 14:29, 16:51, 16:53 and 16:57 Greenwich mean time*

constitutes a highly monochromatic image of the source in the focal plane of the camera. A reflex mirror in the beam permits the image to be examined visually through the eyepiece.

"The operating procedure is relatively simple. After the instrument is assembled and aligned, the angle of the diffraction grating is adjusted to bring the desired spectral line into view in the eyepiece.

"Diffraction gratings produce a series of spectra (spectral orders), an effect analogous to a multiple rainbow. The extent to which the colors are dispersed increases with the 'higher' orders at the cost of brightness. When used in the first order, the grating of my spectrograph can spread 14.5 angstrom units of the spectrum enough to fill the exit slit when its jaws are spaced one millimeter apart. In other words, dispersion in the first order is 14.5 angstroms per millimeter. The second order gives a dispersion of 7.5 angstroms per millimeter, and the succeeding orders proportionately more. Thus, were it not for the fact that the brilliance of the diffracted light diminishes with each successive higher order, one could observe an extremely narrow band of color through an exit slit of substantial width. My grating is ruled for use in the second order (it is 'blazed' for 10,000 angstroms in the first order and 5,000 in the second). Hence, to observe a band of color one angstrom wide the jaws of the exit slit must be spaced about a seventh of a millimeter apart.

"The spectral orders produced by the grating tend to overlap; that is, the red end of the first order falls on the violet end of the second, the red end of the second order overlaps the violet end of the third, and so on. The effect must be minimized or the quality of the final image will suffer. This is accomplished by inserting in the optical path a glass filter which has maximum transmission in the region of the spectrum under observation. If one is observing the alpha line of hydrogen, for example, violet light from the unwanted order will be suppressed by a red filter such as the Corning Glass Works' No. N1661. Similarly, a violet filter is used when observing the H or K lines of calcium. The filter may be inserted at any point in the system, but a filter located at or near the primary focus may heat unevenly and break. Hence the filters are usually placed at a point between six and eight inches from the primary focus.

"With the filter in place, the entrance slit is opened to a width of about .02

inch. This admits considerably more light than is needed for observing but simplifies subsequent adjustments. The exit slit is opened so the spectral lines can be seen easily between the jaws. The desired line is selected and focused sharply by moving the entrance slit back and forth. The image of the exit slit is then focused so that the jaws appear sharp when viewed through the eyepiece. If spectroheliograms are to be made, the camera is similarly ad-

justed to bring the slit into sharp focus on the film. The instrument is next adjusted for maximum resolution. First, the jaws of the entrance slit are closed to the point where the spectral lines appear dark and sharply defined against a light background. Then the jaws of the exit slit are closed until they just frame the line selected. In the case of the alpha line of hydrogen the optimum width will be approximately a fifth of a millimeter. The motor is started. As

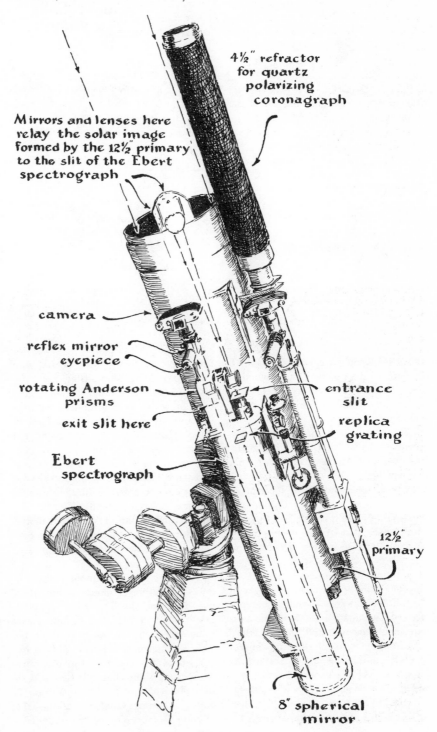

*Semerau's telescope, showing the optical path used during spectroheliograph observations*

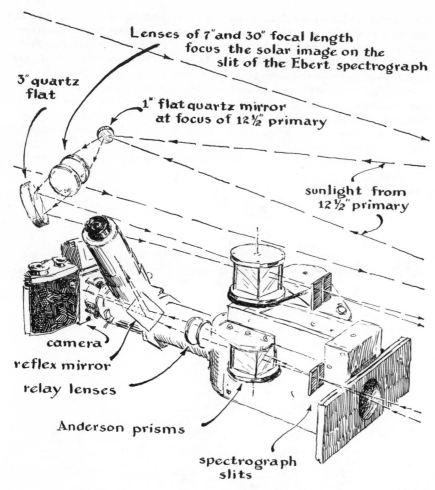

Lenses of 7" and 30" focal length focus the solar image on the slit of the Ebert spectrograph

3" quartz flat

1" flat quartz mirror at focus of 12½" primary

sunlight from 12½" primary

camera
reflex mirror
relay lenses

Anderson prisms

spectrograph slits

*Details of Semerau's spectroheliograph*

the prisms reach a speed of about 16 revolutions per second, a monochromatic image of the sun, complete with the flaming detail of the solar surface, will come into view.

"This of course assumes that all adjustments have been carefully made. Each element of the instrument, from the objective to the eyepiece and camera, must be aligned with the optical axis of the system. If the telescope and spectrograph are out of line, for example, only part of the light will fall on the diffraction grating. The final image will not be as bright as it could be. In addition, the unused light will be reflected from the housing, will mix with the diffracted rays and reduce the contrast of the image. Similarly the system should be adjusted so that white light from the entrance slit approximately fills the diffraction grating. If the grating is not fully illuminated, its efficiency suffers. Conversely, rays which extend beyond the edge of the grating are lost to the final image and impair its contrast.

"The final adjustment consists in gradually narrowing the exit slit. This

brings progressively finer details into view; prominences, mottling near the region of sunspots, dark filaments, flocculi and so on. It also reduces the brilliance of the image and sets a limit to visual observation. At this point the camera comes into use. The average exposure time is from two to four seconds; the instrument is guided during a time exposure as it is in conventional astronomical photography. The camera is also used in the violet region of the spectrum beyond the range of the eye. In this region lie the H and K lines of calcium.

"Although the instrument has many desirable features, I should also mention one disadvantage. The Ebert spectrograph, as I constructed it, introduces some distortion; the image of the sun's disk is somewhat elliptical. This is explained by the fact that the slits must be located somewhat off the axis of the spherical mirror. The distortion is partially compensated by tilting the camera. Curved slits would provide a better correction, but I have no way of making them. The distortion does not impair resolution but it introduces some

complication in locating details accurately on the image. Advantages of the design include simplicity, lightness, relatively low cost and a cylindrical form that is easy to assemble on an equatorial mounting. In addition, desired portions of the spectrum can be brought into view at the twist of a dial.

"The spectroheliograph is mounted beside the coronagraph previously described in 'The Amateur Scientist' [September, 1955]. The two are used simultaneously. The coronagraph shows prominences at the edge of the sun in great detail, but gives no hint of the solar disturbances responsible for them because the central disk is masked by a diaphragm. In contrast, the spectroheliograph reveals faculae, flocculi, filaments, spots and even prominences of exceptional brilliance.

"Amateurs often ask which instrument I prefer. The choice is difficult. Both were interesting projects. The cost is influenced, particularly in the case of the coronagraph, by the extent to which the final image is made monochromatic. The filtering element in my coronagraph is a quartz monochromator, a multi-layered sandwich of crystal quartz and Polaroid film. It was designed to transmit a band of color four angstroms wide. At current prices the raw crystalline quartz from which it was made would cost about $150. Four times this amount would be needed to narrow the pass-band to one angstrom. If, for example, a monochromator designed for a pass-band of four angstroms requires a stack of crystal slabs four inches high, one for a pass-band of one angstrom would require a 16-inch stack. Moreover, each successive slab in the stack must be twice as thick as its predecessor. This means that the final slab in a 16-inch stack would have to be cut from a raw crystal more than eight inches long. Crystals of this size—and of the necessary optical perfection—are not common in nature, and are priced accordingly.

"Monochromators are not easy to build. I would rather make two sets of prisms for a spectroheliograph than any two quartz slabs for the monochromator. Not only is glass softer and easier to work than quartz, but the prisms may be cut in random directions from any location in a block of glass. Quartz crystals must be put through a complex series of tests to determine their optical properties in advance of cutting. The defects of quartz are many, and the substance is so hard that a diamond-edged saw is almost a necessity.

"But one look through the eyepiece

of a coronagraph, even one with a pass-band of four angstroms, justifies the investment of time and labor. When they are seen in detail, solar prominences are among nature's most impressive spectacles. I did not keep an accurate record of my cash outlay for the two instruments, but an estimate of $300 for each would not be far wrong.

"I had the good fortune to observe and photograph an interesting pair of solar events on October 20, 1957. No outstanding disturbance was evident when I began to observe at 14:15 Greenwich mean time, but within 15 minutes a scarlet flocculus appeared near the southwest edge of the sun. The intensity of the flocculus remained constant during the following two hours, but at 16:51 a small flare brightened near the east edge. At about this time the cloud first observed also started to brighten; thereafter both regions grew in size and brightness to the International Geophysical Year standard of 'Importance 3.' By 17:15 the east flare had diminished to normal brightness and 45 minutes later the one near the southwest edge similarly faded. The visual image was sharp and crisp. Poor seeing caused some blurring of the photographs, but conditions improved somewhat at 16:51 [*see photographs on page 132*]."

Semerau states that he is now working on an electronic servo guiding-mechanism, two 35-millimeter time-lapse cameras and a heavier equatorial mounting for his instruments. What he hangs on the mounting next is anybody's guess!

# 26

# Spectrohelioscope

*A new kind of spectrohelioscope to observe solar prominences*

March 1974

Few natural spectacles compare in splendor with the glowing prominences that rise from the surface of the sun. Apart from astronomers few people ever see the display, which is usually lost in the sun's white glare. Those fortunate enough to be in the path of a total eclipse can observe the prominences as scarlet plumes that stand out in vivid contrast to the pearly background of the glowing corona.

Gene F. Frazier of 2705 Gaither Street S.E., Washington, D.C. 20031, views the spectacle routinely with a homemade instrument that in effect blocks from the eye light of all colors except the one emitted with maximum brilliance by the prominences. The hue approximates the darkest red of the setting sun. The emission is radiated by glowing hydrogen at a wavelength of 6,562.8 angstroms. In certain respects Frazier's apparatus resembles the spectrohelioscope previously described in these columns [see "The Amateur Scientist," SCIENTIFIC AMERICAN, April, 1958]. His instrument has an additional diffraction grating but requires no solar telescope or motor-driven optical parts. He describes the principles, construction and operation of the apparatus as follows.

"The effect that my instrument is based on (and that led to the development of the spectrohelioscope) was first described by the French astronomer Pierre Jules César Janssen following his observation in 1868 of the total solar eclipse in India. When Janssen focused the edge of the sun's image on the slit of his spectroscope, he was astonished by the brilliance of the spectral line at 6,562.8 angstroms. It was so bright that on the following day Janssen looked for the color in full sunlight. By opening the

slit of the spectroscope he discovered that he could observe a portion of the prominence. A few days later the British astronomer Joseph Norman Lockyer hit on the same technique without the benefit of an eclipse.

"News of the discovery fascinated amateurs of the day, primarily because the brightness of the sun made observations possible with instruments of small aperture and proportionately low cost. It turned out, however, that home-built spectroscopes scattered too much light for satisfactory results. In addition clockwork-driven structures capable of keeping an image of the sun's edge focused exactly on the thin slit of the spectroscope called for a higher level of craftsmanship than most amateurs could attain.

"The design of my instrument sidesteps these requirements. Essentially the instrument employs an external diffraction grating to disperse and reflect sunlight to a concave mirror [see illustration on opposite page]. The mirror projects the rays through an adjustable plate of flat glass to a focus in the plane of the entrance slit of a conventional spectroscope.

"A photograph that could be made by putting a photosensitive plate in the position occupied by the slit of the spectroscope would not show the dark absorption lines that normally characterize the solar spectrum. In my system the image of the sun functions as the slit. Hence a photograph is composed not of the series of narrow absorption lines but of overlapping images of the solar disk separated by distances corresponding to the wavelength of the absorbed light.

"The adjustable plate of flat glass that admits incoming light to the slit acts as a vernier for displacing the rays laterally

with respect to the slit. Rays that enter the plate at an angle to its perpendicular are refracted and emerge at the identical angle. The amount of deviation is approximately proportional to the angle between the plate and the entering beam. By rotating the plate the observer can shift the spectrum any small distance with respect to the slit. The plate functions as a precision tuner that enables the experimenter to admit any narrow portion of the spectrum to the slit.

"The selected rays, which may span a range of color only 10 angstroms wide, emerge from the slit as a diverging beam. The diverging rays fall on a concave mirror from which they are reflected as a bundle of parallel rays to the internal diffraction grating of the spectroscope [see illustration on pages 138 and 139]. The internal grating disperses the colors still more. The angle at which the internal grating is set can be adjusted to reflect rays of essentially monochromatic light to the second concave mirror of the spectroscope. The second mirror reflects the rays to focus in the plane of the eyepiece.

"The details of the filtering action can be demonstrated by replacing the external grating with a flat mirror and letting sunlight fall on the mirror. After adjustment an instrument so modified would display at the eyepiece the normal solar spectrum crossed by dark absorption lines. Assume that the geometry of the diffraction grating of the spectroscope is such that each angstrom of wavelength of the solar spectrum is dispersed through a distance of one millimeter in the focal plane of the eyepiece. This was essentially the case with Janssen's spectroscope. The chromosphere of his solar image was less than one millimeter wide. Therefore he could partly isolate the emission of the prominences from background light by carefully focusing this narrow feature of the image on the slit of his spectroscope.

"Now assume that the flat mirror is replaced by the external diffraction grating of my instrument and that the angle of the grating is carefully adjusted to reflect a narrow band of light on the slit that spans 10 angstroms (from, say, 6,558 to 6,568 angstroms). The spectrum is noncoherent. For this reason the light that reaches the slit consists of many monochromatic images of the sun's disk that overlap on each side of the hydrogen line at 6,562.8 angstroms. If the dispersion of the gratings is assumed to be one angstrom per millimeter, the centers of each of the images of the sun's disk would be separated by one milli-meter. At any setting 10 solar disks would overlap.

"This means that a band of color only 10 angstroms wide can enter the slit and that the scattering of light is significantly reduced. When the instrument is adjusted for observing prominences at 6,562.8 angstroms, unwanted light is reduced by more than 95 percent! Indeed, on a clear day it is not unusual for the field to appear completely dark five angstroms from the image. The full solar image appears in the field of view, which helps the observer to keep the edge of the image centered on the 6,562.8-angstrom line. With the aid of the tuner I have easily observed prominences continuously for intervals of more than two minutes.

"The construction requires no special skills, but the quality of the gratings is crucial. They must be mounted with care. The gratings should be ruled with at least 1,200 lines per millimeter for adequate resolution and high dispersion. The ruled area of the gratings in my instrument measures two inches square. The lines are blazed for 6,600 angstroms in the third order, which is to say that the surface of the rulings is cut at an angle that reflects light of maximum intensity at the 6,600-angstrom wavelength in the same direction as the grating disperses these rays in the third order.

"The gratings can be mounted in simple structures bent in the form of a V from sheet steel or brass. The gratings can be attached lightly to these mountings with machine screws and insulated

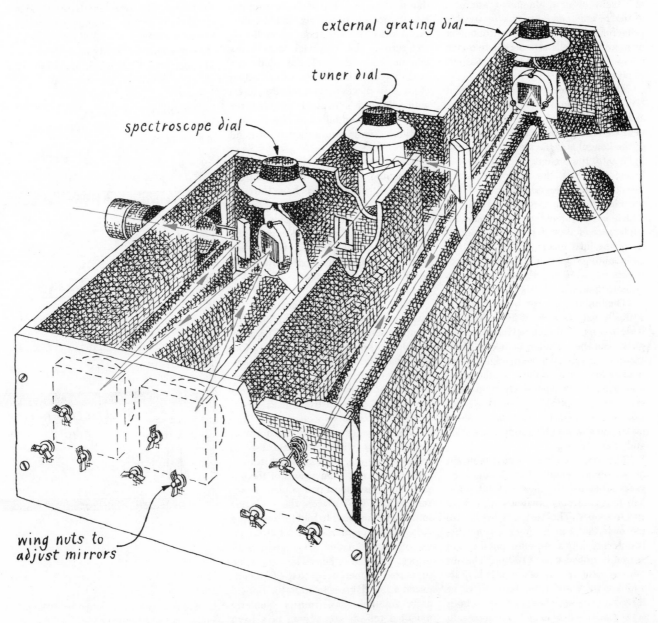

*Double-grating filter of Gene F. Frazier's spectrohelioscope*

from the metal with felt lining. The metal V's are supported by soldering the rear side to a quarter-inch copper rod that fits a radio dial of the vernier type [see top illustration on page 140]. The copper rod must be bent to an angle such that the projected axis of the vernier dial bisects the plane of the grating. When the rod is so mounted, the angle of the gratings with respect to the impinging rays can be altered without displacing the spectral orders at the eyepiece.

"I made adjustable cells of plywood for supporting the mirrors. The cells are supported at three equidistant points by machine screws fitted with compression springs and wing nuts. The mirrors can be lightly fastened to the plywood with wood screws insulated by rubber tubing and fiber washers. Incidentally, adjustable cells of cast aluminum are now available commercially at a reasonable price for mirrors of three-inch diameter or more. The cells also accept two-inch mirrors mounted in three-inch washers of plywood.

"The diameter of the mirrors is not critical, but it must be at least as large as the ruled area of the gratings to prevent vignetting (obscuration at the edges of the image) and the scattering of stray light into the image. In addition the focal length of the objective mirror should be an integral multiple of the focal length of the spectroscope mirrors, which, in turn, should be equal to within a tolerance of about 2 percent. The quality of the final image can be optimized by mounting the two mirrors of the spectroscope as close together as possible in order to minimize off-axis aberrations.

"The tuning plate can be made of any optically flat glass about 50 millimeters wide and six to 10 millimeters thick. The piece can be circular or rectangular. Plates of this size that were originally intended for use as optical windows are now available inexpensively from dealers in surplus optical supplies. The plate can be mounted by a frame of sheet metal and adjusted by a supporting shaft and a vernier dial.

"The construction of an adjustable slit of adequate quality has through the years remained the most difficult problem that confronts amateurs who make spectroscopes. The best slits by far are the ones that can be bought from distributors of optical supplies, but they are currently priced from $100 up. The slit must remain rigidly centered when its width is increased from zero to two or three millimeters, which means that both jaws must move equal distances in opposite directions when the device is adjusted.

"A mechanical linkage in the form of a parallelogram can satisfy this condition [see illustration at bottom left on page 140]. The system of links can be assembled with snugly fitting machine screws. Excess play in the system can be eliminated by inserting a pair of helical springs to maintain a few grams of tension between the side links that support the jaws of the slit. The jaws can be made of single-edge safety-razor blades, which can be fastened to the supporting links with epoxy cement.

"I prefer jaws made of sections cut from a hacksaw blade. I first grind off the saw teeth with a carborundum wheel. The opposite edges are polished to remove surface irregularities. The procedure is not difficult. I grind two four-inch slabs of plate glass together with a thin slurry of No. 120 carborundum grit in water for a period of six minutes, making elliptical strokes about an inch long and turning the 'sandwich' over every minute. The edges of the blade are ground against the frosted side of either of the glass pieces for two minutes, again with elliptical strokes. I examine the edges for pits and hills and, if necessary, continue grinding until they are straight and smooth.

"One of the completed jaws is soldered or cemented with epoxy to its supporting linkage. After the jaw has been fastened the linkage is moved to the position where the separation of the side links is at a minimum. The ground edge of the companion jaw is placed in full contact with the ground edge of the jaw previously installed and similarly attached to its supporting link.

"All optical elements of the instrument must be installed in a lightproof housing. I improvised one of plywood. The spectroscope was made as a separate unit that could be bolted to the housing that supports the external grating and the objective mirror. This arrangement enables me to employ the instrument as a conventional laboratory spectroscope.

"The housings can take the form of simple boxes with removable lids to which the vernier dials are fixed. I minimized the overall dimensions of the apparatus by inserting a pair of plane front-surface mirrors between the objective mirror and the tuner to fold the incoming rays. The plane mirrors are set at an angle of approximately 90 degrees with respect to each other. They must be supported by mountings that can be adjusted a few degrees to align the optical path. Another plane mirror similarly mounted reflects converging rays from the spectroscope to the focal plane of the eyepiece. The interior of the spectroscope housing, including particularly the barrier that separates the concave mirrors, should be painted flat black to minimize the reflection of scattered light.

"All optical parts should be tested before they are installed. Testing the focal length of the mirrors is particularly important. A simple measurement of the focal length can be made by standing the mirror on edge, directing a flashlight toward the metallized surface and catching the reflected image of the lamp filament on a sheet of white paper placed next to the flashlight. Vary the distance of the flashlight and the paper from the mirror until the sharpest possible image of the lamp filament appears on the paper. The focal length of the mirror is equal to exactly half the distance between the image of the filament and the surface of the mirror. The focal length can be measured with greater accuracy by setting up the knife-edge test used for checking telescope mirrors [see Amateur Telescope Making: Book One, edited by Albert G.

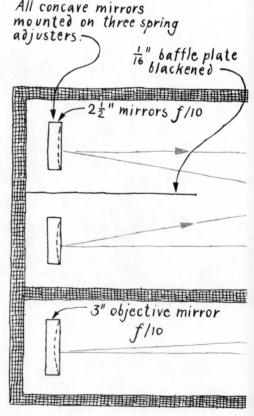

All concave mirrors mounted on three spring adjusters.

$\frac{1}{16}$" baffle plate blackened

2$\frac{1}{2}$" mirrors f/10

3" objective mirror f/10

Ingalls, Scientific American, Inc.].

"An instrument similar to mine can be built for less than $250. The cost will increase exponentially with the diameter of the optical parts. The two-inch gratings and mirrors have enabled me to make most of the observations I had in mind when I undertook the construction and also to do a variety of laboratory experiments.

"The instrument has a maximum dispersion of about two angstroms per millimeter, which is equivalent to displaying the rainbow as an image more than six feet wide at the focal plane of the eyepiece. The zone in which the prominences can be viewed at the edge of the sun is less than one millimeter wide, but even so it is sufficient to enable the observer to isolate the 6,562.8-angstrom spectral line of hydrogen.

"When the apparatus has been attached to an equatorial mounting that includes a clock drive to keep the ruled surface of the grating pointed approximately toward the sun, set the tuning plate at a right angle to the optical path and start the clock drive. Cover the objective mirror with a disk of white paper and adjust the external grating to the angle at which reddish light from the sun falls on the paper. Transfer the paper to the position of the slit of the spectroscope and adjust the flat mirrors to angles such that the reddish image falls on the paper at the position of the slit.

"Remove the paper. Adjust the first mirror of the spectroscope (the collimating mirror) to the angle at which the now parallel rays of the reddish light flood the diffraction grating of the spectroscope. If the light is difficult to see, cover the grating with a small sheet of white paper. Adjust the second mirror of the spectroscope to project converging rays to the flat mirror adjacent to the eyepiece. A sheet of paper inserted in the focal plane of the eyepiece should now display an image of the sun.

"While viewing through the eyepiece adjust the angle of both gratings to center the dark 6,562.8-angstrom spectral line on the slit of the spectroscope. The line is the darkest and broadest one in this region of the spectrum. The observer should now see in the eyepiece the complete scarlet image of the sun.

"To observe the limb, rotate the tuner so that the image appears to shift along the absorption spectrum to the point at which the edge of the sun is centered on the line at 6,562.8 angstroms. That is the adjustment at which Janssen made his initial observation. If a prominence happens to be located at this point, the observer will see the absorption line fade and be replaced by a bright area.

"Only a small area of the sun's edge is being observed. For this reason the edge will appear to curve away from the straight slit. To see prominences along a substantial portion of the edge the straight slit must be replaced by one that matches the curvature of the image.

"Curved slits of two kinds are relatively easy to make. Of the two I prefer one that requires the use of an engine lathe to cut a disk of metal equal in diameter to the diameter of the solar image at the focal plane of the objective mirror. The dimension is very nearly

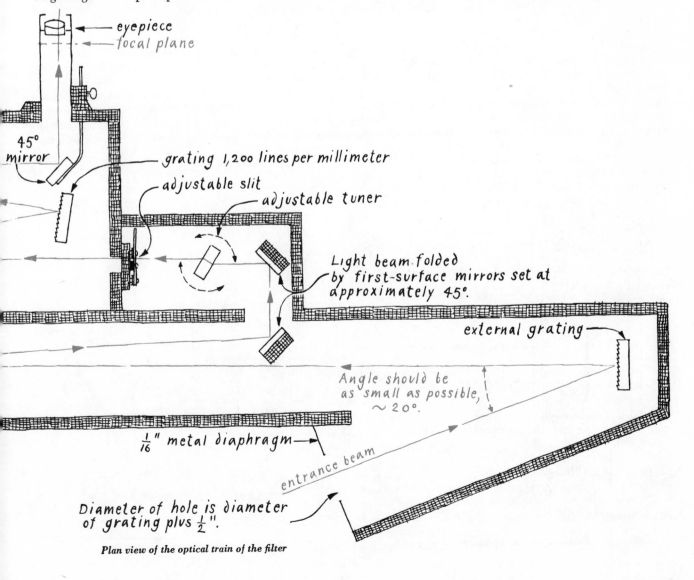

eyepiece
focal plane

45°
mirror

grating 1,200 lines per millimeter
adjustable slit
adjustable tuner

Light beam folded
by first-surface mirrors set at
approximately 45°.

external grating

Angle should be
as small as possible,
~ 20°.

$\frac{1}{16}$" metal diaphragm

entrance beam

Diameter of hole is diameter
of grating plus $\frac{1}{2}$".

*Plan view of the optical train of the filter*

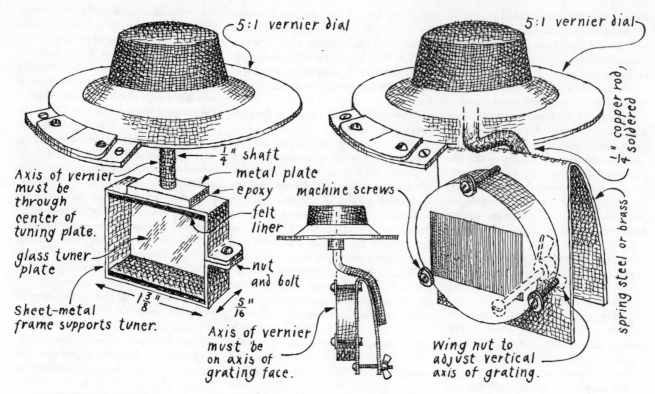

*Vernier mounting of the tuning plate* (left) *and the diffraction gratings* (right)

equal to the focal length of the objective divided by 114.

"In the case of an objective mirror with a focal length of 30 inches the radius of the curved slit is approximately .131 inch. A hole .05 inch larger in radius is drilled in a metal sheet. The device is assembled by centering the disk in the hole and tacking it in place with a dab of solder or epoxy cement to leave a clear arc .05 inch wide and extending about 180 degrees.

"Another technique for making the curved slit is easier. Drill a hole slightly larger in diameter than the solar image and place it over a mirror that has been aluminized or silvered. Dip the sharpened tip of a wood toothpick in dilute nitric acid. Shake excess acid from the wood. Insert the sharpened tip through the hole in the metal and, with the metal

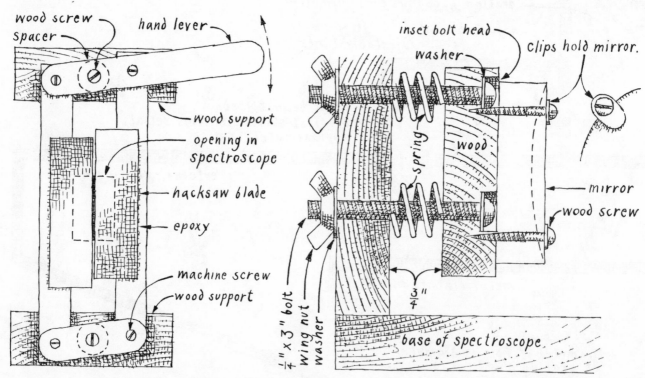

*Details of the slit mechanism*

*Adjustable cells for concave mirrors*

as a guide, trace a semicircle on the coated glass. If the operation is performed with care, the acid will etch a usable slit in the metallic film. To observe solar prominences substitute the curved slit for the straight one.

"In a sense the double-grating filter is analogous to the tuning system of a radio set. It enables the investigator to select for observation a narrow band of light waves much as the dial of a radio set tunes in a narrow interval of the radio spectrum. The filter can serve in a variety of experiments other than the observation of solar prominences. For example, my interests include the use of polarimetry for investigating the characteristics of minerals. At various orientations the surface of a rock can be seen in different colors, which depend on the crystal structure of the specimen. By examining the rock with the tuned filter, with both gratings adjusted to an appropriate angle, distinct colors appear in various areas of the surface that characterize the specimen.

"Mineralogists also routinely dissolve bits of unknown rock in a bead of incandescent borax to identify its constituents by the characteristic colors of the resulting flame. Each chemical element radiates a unique set of wavelengths. With the double-grating filter the experimenter can observe and even photograph the distribution of elements in the glowing gases.

"The rulings of a grating are cut at an angle that optimizes the efficiency of the device as a reflector of light of specified wavelength at a specified angle with respect to the plane of the rulings. As I have mentioned, the angle at which the rulings are cut is called the blaze. The angles at which gratings reflect bundles of rays dispersed in the form of spectra are known as the spectral orders.

"As I have mentioned, the gratings of my instrument are blazed to reflect most of the incident light at 6,600 angstroms in the third order. In general dispersion increases with the spectral order at the cost of brightness. My gratings were selected primarily for viewing solar prominences. Hence they were blazed for maximum brightness of the deep red in the spectral order that resulted in a dispersion of two angstroms per millimeter at the focal plane of the eyepiece. People who design the instrument for experiments of other kinds such as flame spectroscopy would doubtless select gratings blazed for other colors in other spectral orders.

"All optical parts from which the instrument is made are available from the Edmund Scientific Co. (300 Edscorp Building, Barrington, N.J. 08007). Diffraction gratings of the exact kind in my instrument are distributed by Jarrell-Ash Division of the Fisher Scientific Co. (590 Lincoln Street, Waltham, Mass. 02154) and by Bausch & Lomb Inc. (526–68 Lomb Park, Rochester, N.Y. 14602).

"Two people have helped me with this study. Timothy O'Hover of the chemistry department of the University of Maryland provided a laboratory and Gary A. Frazier supplied inspiration and electronic testing equipment for the original studies. I am deeply grateful to them."

# BIBLIOGRAPHIES

## I. LASERS

### 1. Helium-Neon Laser

THE LASER. A. Yariv and J. P. Gordon in *Proceedings of the IEEE*, Vol. 51, No. 1, pages 4–29; January, 1963.

OPTICAL MASERS. Arthur L. Schawlow in *Scientific American*, Vol. 204, No. 6, pages 52–61; June, 1961.

PROCEDURES IN EXPERIMENTAL PHYSICS. John Strong, Roger Hayward, H. Victor Neher, Albert E. Whitford and C. Hawley Cartwright. Prentice-Hall, Inc., 1938.

### 2. More on the Helium-Neon Laser

EXPERIMENTS IN PHYSICAL OPTICS USING CONTINUOUS LASER LIGHT. T. J. Perkins. Optics Technology, Inc., 1964.

### 3. Argon Ion Laser

FUNDAMENTALS OF OPTICS. Francis A. Jenkins and Harvey E. White, McGraw-Hill Book Company, Inc., 1950.

INTRODUCTION TO LASER PHYSICS. Bela A. Lengyel. John Wiley & Sons, Inc., 1966.

CREATIVE GLASS BLOWING. James E. Hammesfahr and Clair L. Stong. W. H. Freeman and Company, 1968.

### 4. Tunable Dye Laser

FLASHLAMP-PUMPED ORGANIC-DYE LASERS. P. P. Sorokin, J. R. Lankard, V. L. Moruzzi and E. C. Hammond in *The Journal of Chemical Physics*, Vol. 48, No. 10, pages 4726–4741; May 15, 1968.

### 5. Carbon Dioxide Laser

AMATEUR TELESCOPE MAKING: BOOK ONE. Edited by Albert G. Ingalls. Scientific American Incorporated, 1950.

HIGH-POWER CARBON DIOXIDE LASERS. C. K. N. Patel in *Scientific American*, Vol. 219, No. 2, pages 22–33; August, 1968.

### 6. Infrared Diode Laser

PROCEDURES IN EXPERIMENTAL PHYSICS. John Strong in collaboration with H. Victor Neher, Albert E. Whitford, C. Hawley Cartwright and Roger Hayward. Prentice-Hall Inc., 1938.

GALLIUM ARSENIDE LASERS. Edited by C. H. Gooch. John Wiley & Sons, Inc., 1969.

A NEW CLASS OF DIODE LASERS. Morton B. Panish and Izuo Hayashi in *Scientific American*, Vol. 225, No, 1, pages 32–40; July, 1971.

### 7. Nitrogen Laser

A SIMPLE PULSED NITROGEN 3371 A LASER WITH A MODIFIED BLUMLEIN EXCITATION METHOD. J. G. Small and R. Ashari in *The Review of Scientific Instruments*, Vol. 43, No. 8, pages 1205–1206; August, 1972.

## II. HOLOGRAMS

### 8. Homemade Holograms

FUNDAMENTALS OF OPTICS. Francis A. Jenkins and Harvey E. White. McGraw-Hill Book Company, Inc., 1950.

A NEW MICROSCOPIC PRINCIPLE. D. Gabor in *Nature*, Vol. 161, No. 4098, pages 777–778; May 15, 1948.

### 9. Insuring Stability of the Apparatus

COFFEE-TABLE HOLOGRAPHY. John Landry in *Journal of the Optical Society of America*, Vol. 56, No. 8, page 1133; August, 1966.

HOLOGRAPHY FOR THE SOPHOMORE LABORATORY. Robert H. Webb in *American Journal of Physics*, Vol. 36, No. 1, pages 62–63; January, 1968.

### 10. Holograms with Sound and Radio Waves

X-BAND HOLOGRAPHY. R. P. Dooley in *Proceedings of the IEEE*, Vol. 53, No. 11, pages 1733–1735; November, 1965.

A METHOD FOR PHOTOGRAPHING MICROWAVE WITH A POLAROID FILM. Keigo Iizuka. Harvard University Technical Report No. 558. March, 1968.

OPTICAL FILM SENSORS FOR RF HOLOGRAPHY. H. E. Stockman and Berthold Zarwyn in *Proceedings of the IEEE*, Vol. 56, No. 4, page 763; April, 1968.

## III. INTERFEROMETERS

### 11. Michelson Interferometer

INEXPENSIVE MICHELSON INTERFEROMETER. Eric F. Cave and Louis V. Holroyd in *American Journal of Physics*, Vol. 23, No. 1, pages 61–63; January, 1955.

### 13. Speckle Interferometer

LASERS: GENERATION OF LIGHT BY STIMULATED EMISSION. Bela A. Lengyel. John Wiley & Sons, Inc., 1971.

### 14. Series Interferometer

INTERFEROMETERS. J. Dyson in *Concepts of Classical Optics* by John Strong, W. H. Freeman and Company, 1958.

### 16. Interferometer to Measure Dirt Content of Water

THE EFFECT OF LIGHT ON THE SETTLING OF SUSPENSIONS. C. G. T. Morison in *Proceedings of the Royal Society of London*, Series A, Vol. 108, No. A746, pages 280–284; June 2, 1925.

TURBIDIMETRY AND NEPHELOMETRY in *Encyclopedia of Chemical Technology*: Vol. XX, edited by R. E. Kirk and D. F. Othmer. Wiley-Interscience, 1969.

## IV.   INSTRUMENTS OF DISPERSION

### 17. Ocular Spectroscope

EXPERIMENTAL SPECTROSCOPY. Ralph A. Sawyer. Prentice-Hall, Inc., 1944.

CHEMICAL SPECTROSCOPY. Wallace R. Brode. John Wiley & Sons, Inc., 1943.

### 18. Bunsen Spectroscope and Note on Making Liquid Prisms

EXPERIMENTAL SPECTROSCOPY. Ralph A. Sawyer. Prentice-Hall, Inc., 1944.

### 19. Diffraction-Grating Spectrograph

EXPERIMENTAL SPECTROSCOPY. Ralph A. Sawyer. Prentice-Hall, Inc., 1944.

TELESCOPES AND ACCESSORIES. George Z. Dimitroff and James G. Baker. The Blakiston Company, 1945.

### 20. Diffraction-Grating Spectrograph to Observe Auroras

THE AMATEUR SCIENTIST. C. L. Stong. Simon and Schuster, Inc., 1960.

THE AURORAE. L. Harang. John Wiley & Sons, Inc., 1951.

EXPERIMENTAL SPECTROSCOPY. R. A. Sawyer. Prentice-Hall, Inc., 1944.

### 21. Inexpensive Diffraction-Grating Spectrograph

CHEMICAL SPECTROSCOPY. W. R. Brode. John Wiley & Sons, Inc., 1943.

### 22. Ultraviolet Spectrograph

FUNDAMENTALS OF OPTICS. Francis A. Jenkins and Harvey E. White. McGraw-Hill Book Company, Inc., 1950.

### 23. Inexpensive Spectrophotometer

ANALYTICAL ABSORPTION SPECTROSCOPY: ABSORPTIMETRY AND COLORIMETRY. Edited by M. G. Mellon. John Wiley & Sons, Inc., 1950.

CHEMICAL SPECTROSCOPY. Wallace R. Brode. John Wiley & Sons, Inc., 1943.

OPTICAL METHODS OF CHEMICAL ANALYSIS. Thomas R. P. Gibb, Jr. McGraw-Hill Book Company, Inc., 1942.

### 24. Recording Spectrophotometer

ANALYTICAL ABSORPTION SPECTROSCOPY: ANASORPTIMETRY AND COLORIMETRY. Edited by M. G. Mellon. John Wiley & Sons, Inc., 1950.

### 25. Spectroheliograph

AMATEUR TELESCOPE MAKING—BOOK ONE. Edited by Albert G. Ingalls. Scientific American, Inc., 1957.

AMATEUR TELESCOPE MAKING—ADVANCED. Edited by Albert G. Ingalls. Scientific American, Inc., 1957.

AMATEUR TELESCOPE MAKING—BOOK THREE. Edited by Albert G. Ingalls. Scientific American, Inc., 1956.

### 26. Spectrohelioscope

FUNDAMENTALS OF OPTICS. Francis A. Jenkins and Harvey E. White. McGraw-Hill Book Company, Inc., 1957.

TOOLS OF THE ASTRONOMER. Gerhard R. Miczaika and William M. Sinton. Harvard University Press, 1961.

# INDEX